Exclusive signed edition

Anna May McHugh (signature)

ANNA MAY MCHUGH

September 2017

Queen of the Ploughing

Queen of the Ploughing

My Story

ANNA MAY McHUGH

with the assistance of Alison Healy

PENGUIN

IRELAND

PENGUIN IRELAND

UK | USA | Canada | Ireland | Australia
India | New Zealand | South Africa

Penguin Ireland is part of the Penguin Random House group of companies
whose addresses can be found at global.penguinrandomhouse.com.

First published 2017
001

Copyright © Anna May McHugh, 2017

The moral right of the author has been asserted

The author and publisher would like to thank Pat MacConnell for permission to
include 'Ballad of Anna May', Mark Bergin for permission to include 'The Song of the Plough'
and Pat Fortune for permission to include a verse from 'Fair Play to You Martin
Kehoe'. The lines from 'Ploughman' by Patrick Kavanagh are reprinted from *Collected
Poems*, edited by Antoinette Quinn (Allen Lane, 2004), by kind permission of the Trustees
of the Estate of the late Katherine B. Kavanagh, through the Jonathan Williams
Literary Agency

Set in 13.5/16 pt Garamond MT Std
Typeset by Jouve (UK), Milton Keynes
Printed in Great Britain by Clays Ltd, St Ives plc

A CIP catalogue record for this book is available from the British Library

ISBN: 978–1–844–88421–6

www.greenpenguin.co.uk

To
The memory of my late husband, John
My son and daughter, DJ and Anna Marie
My grandchildren, Seán Óg, Tadhg, Dearbhla and Saran
My daughter-in-law, Fiona, and son-in-law, Declan
My brothers and sisters

You are all my inspiration

'Turn down the green, O man who ploughs:
Guide thou the plough with sharpened share!
Turn up the brown to sapphire skies!
Mankind on thee for bread relies.'

From 'The Song of the Plough' by JJ Bergin, co-founder
of the National Ploughing Championships

'The Ploughing Championships are a truly Irish
phenomenon, one of the highlights of the year for so
many. They are an opportunity to meet up with old
friends, catch up with the latest farming news, and
enjoy the great atmosphere that makes 'the Ploughing'
such a unique and enduring event . . . these
Championships are truly the Olympics of the land.'

President Michael D. Higgins, opening the 85th
National Ploughing Championships in Screggan,
County Offaly, in 2016

Contents

CONTENTS

1. Scenes from a childhood

My childhood memories are a string of small scenes . . .

> . . . Mam in her crossover apron, head bent over the sewing machine.
> . . . Dad standing still in the farmyard, pausing for the Angelus.
> . . . the joy of waking on Christmas morning to the smell of the pudding that had been simmering all night.
> . . . the shock of brightness at night when the electricity came to our home.
> . . . the mellow sound of my mother's melodeon playing *The Spinning Wheel*.

We were blessed with a truly happy family home, and that was all thanks to Mam and Dad. My story begins in Clonpierce, which is about two miles from the village of Ballylinan in County Laois. We live close to the County Kildare border, and Athy is our nearest town, but I am most definitely a proud Laois woman. So that's where I was born on Friday, 11 May 1934. That's right. I'm eighty-three years young as I write this. I have often dodged the question of my age when I've been interviewed as I'm not the kind of person who shouts their age from the rooftops. That's for reasons I'll explain later. But it would be quite difficult to tell your life story without giving a date of birth, so there you have it.

As it was, 1934 was an eventful year on the world stage, with dictators like Stalin and Mussolini flexing their muscles and Adolf Hitler declaring himself to be the Führer, or leader, in Germany. Cole Porter and Bing Crosby were topping the charts, and the American outlaws Bonnie and Clyde were killed by police in an ambush. But on the very day I was born, something else happened. A massive dust storm, about 1,500 miles long, was spotted moving across America from the Midwest, as far as New York and Boston. It happened because farmers had cut down prairie grasses and ploughed the land to grow crops. When the drought came, there was no prairie grass to hold down the soil and millions of tonnes of topsoil blew across the country. So, you could say that I was inextricably linked with ploughing from the day of my birth.

I was the fifth of eight children, following after Betty, JJ, Paddy and Gerald. Then after me came Stannie, Oliver and Eileen. Similar to many Irish families of that time, we were like steps of stairs, with only a year-and-a-half or so between us all.

My parents did not have an easy time in the early 1930s. They had three young children and a fourth on the way – Gerald – when our house burned down. My father's mother, Annie, also lived with us in the house, as it was her home place. It was a lovely, long thatched house. The chimney went on fire and it was burning away, but no one in the house realized what was happening until a passer-by spotted the smoke and alerted my parents. Mam often spoke with regret about the things that were lost in the fire, but she and Dad also reflected on how lucky they were that they all got out safely. They managed to gather everyone up and escape from the house before the roof collapsed. It must have been awful to have stood helplessly watching their belongings going up in smoke.

They, and Grandmother Annie, moved into my mother's home place up the lane, and Gerald was born soon after, in March 1933. Then they all moved into the loft in one of the farm buildings at Clonpierce while the new house was being built. This is where I, a little red-headed bundle, entered the world, fourteen months after Gerald. I was often told that Betty danced around the kitchen the evening I was born because she had only three brothers to play with and she was delighted to get a baby sister. One month and one day later, we moved into our new house. That loft would later serve as an office for me, before I got married, but more about that later.

The fire must have been an awful setback for my parents because my father had to find the resources to build a new house at a time when his family was expanding rapidly. And what about my poor mother, moving into a new house as she was recovering after giving birth, and with four other young children around her feet? At least she had Grandmother Annie on hand to help with the rearing of the children. Many mothers-in-law lived with their daughters-in-law back then and you would often hear of tensions between them, but Mam and Grandmother Annie had an exceptionally good relationship. Mam's mother died three months before I was born, so I suppose Grandmother Annie became her mother figure.

I was named after Annie, and the May was added in recognition of the month of Our Lady. Grandmother Annie's mother, also Annie, was still alive when I was growing up. Grandmother died first and when we told Great Grandmother that her daughter had died, she said: 'Ah sure, isn't she old?' But my grandmother was only in her early seventies. Great Grandmother must have been in her mid-nineties when she died.

I have to say, I thought the world of my parents. My father, James Brennan, was an only child and he married his next-door neighbour, Elizabeth Wall. She was the youngest of ten children. They were so different and yet so well matched. She was a small, quiet woman while he was a big, broad man. I never heard much about their courtship, but I remember we used to joke with Mam about how they met. She used to say: 'I saw him in a crowd. Sure, you couldn't miss him.'

If people like my mother are not in heaven now, well then the rest of us have no chance at all. I can well remember Mam having to wash the eight of us in a tin bath, and that same bath was used, with a washboard and carbolic soap, for washing the clothes. We had one of those wooden pulley lines over the range that you could lower down to put the clothes on and raise it up again. A great invention, when you think about it.

I often think about the hardship that women like her went through, with big families and no modern conveniences. I can still see her in her crossover apron, starting into baking and mending when we were all going to bed at night. She'd be sitting at her Singer sewing machine, making clothes or mending them or maybe altering them for the younger ones. And she'd still have to be up early in the morning to get us all out to school.

We loved her cooking. She was absolutely wonderful at making éclairs, a thing I would rarely make now because choux pastry is a tricky thing to perfect. Mother was just a naturally good baker. She was very particular about how she made the éclairs, and I can still see her stirring the milk and butter on the old range before mixing in the flour and eggs. If she didn't have cream for the centre of the éclairs, she would use custard. She used to leave them on a wire tray on the windowsill to cool. One day, one of the boys came in and said the cat had taken the éclairs. You wouldn't believe the speed of Mother as she shot out the door after the cat. He was only joking, of course.

She used to have the big bag of flour in a flour bin and it was dipped into every day. Like most women of her time, she baked a lot of brown and white bread. Great big cartwheels of bread because there were a lot of mouths to feed. And she used to make a lovely big currant cake. Now the fruit might have been as rare as hen's teeth in it, but she did her best with what she had. It was a real treat to get two slices of currant cake going to school, along with the bottle of milk. Fried bread was another great treat for us. Even to this day I love to get a piece of buttered bread and toast it in the pan. Thinking about it now makes me hungry.

When we got Mam and Dad gone to town for a good stretch of time, we'd get out the sugar and butter and make butterscotch. This was a real treat for us because we got very few treats back then, and there was no such thing as pocket money to buy sweets. Making butterscotch was a delicate process. If we heated the sugar too quickly, there would be a burnt taste off the butterscotch. We girls wouldn't trust the boys to stir the mixture for that reason. I remember waiting patiently for the mixture to harden. I think the boys would almost smell when it was set and they'd arrive to sample the produce. The most essential thing was to clean the saucepan well and leave no trace, or else Mam would know what we had been up to. Cleaning the saucepan was definitely not part of the fun because there was no washing-up liquid to help remove the remnants of the mixture that were welded to the sides of the saucepan.

*

The day I started school in St Patrick's National School, Ballylinan, gave me my first clear memory. It was nothing dramatic. I have a picture in my head of me, sitting on the knee of my teacher, Mrs Fleckney, our Baby Infants teacher. I suppose I was whingy and she was comforting me. Mrs Fleckney was a lovely, soft, gentle teacher and very kind to the little ones. Starting school was such a wrench for us because we had no playschool to get us acclimatized to big school. Not that we were in the centre of a metropolis. Ballylinan had a four-teacher school at that time, but it still seemed like a very big, scary place to a five- or six-year-old. That school was built in 1934 because Ballylinan had outgrown the old school, but the local population eventually outgrew my school too and a new one was opened in 2013.

We had those old school desks that are now seen as decorative vintage pieces. They had built-in inkwells and you could lift the table-top and store your copy and nib pen in the drawer part. Your seat was attached to the table.

Like most children of the time, we were slapped in school, and the teachers didn't hold back. We were slapped with a ruler if our homework wasn't right, or if we were caught talking, or if we were late for school. And very often children would be late for a genuine reason. They might be held back to help feed calves or milk cows, but the teachers didn't want to hear about it. It didn't take much to get the ruler. One teacher had a phobia about yawning. If she caught you yawning, she would really chastize you because she thought you'd set off a chain of yawning in the classroom. As if a few children yawning would be the end of the world!

But the boys most certainly got it worse than us and it wasn't the ruler they got – it was the stick. I was in the same room as my brother Gerald when I was in fifth class – he was in sixth – and I truly felt for him when I saw him being hit. There was no such thing as parents coming down to the school to complain when this happened. In fact, you didn't tell your parents if you could avoid it because that would only land you in more trouble.

Some people would say the punishment didn't do us any harm, that we needed it at the time. It didn't do me any harm, but some children got it far worse than I did, and who knows the damage it did to them? My brothers would still talk today about the slaps they got, even though it's well over fifty years since they left school. They haven't forgotten how bad it was, so it definitely affected some people.

Children nowadays are much happier going to school than we were. There's a whole different atmosphere between the children, teachers and parents. I think it's great the way teachers get children to make things for their parents on days like Mother's Day and Father's Day. It brings out the best in the child and makes them enjoy school. Parents are more involved in the school now, which is good. I have no memory of my parents ever being invited to go near the place to see how we were getting on.

But I have some great memories of school too, of meeting my friends and playing at lunchtime. Hide-and-seek was one of our favourite games. We loved hopscotch too, and putting markers on the ground and having running races. Oh, the simplicity of it. Being selected to do a job for the teacher was also a special occasion – jobs like ringing the break bell or carrying a message from your teacher to another. When we were in the senior classes, we might even be asked to run up the village to get a message for the teacher from the shop. That was a great honour indeed.

I remember the diocesan examiner coming to the school

every October and we always got new clothes for that. The teachers would be very worried about that visit and they'd be really anxious, looking out the window all morning as they waited for the examiner to arrive. The children who would be best able to answer questions would be put up to the top of the class. I never had that privilege! Nor did I want it. It wasn't all work at school, though. We were taught a bit of Irish dancing and myself, Maureen Clandillon and Doreen Timpson would do the three-hand reel and jig if there was a special occasion.

Mrs Tobin had a little shop beside the school and it was a big treat to get a penny bar. And if you got a KitKat, well, you were made up. I can tell you, there were no niceties thrown into our schoolbags back then. No Friday treats for the lunch-box. In fact, we didn't even have a lunchbox – the lunch was wrapped in a piece of brown paper and you got a bottle of milk, too. I guess the fact that we never ate processed food was why obesity and diabetes were rarely heard of.

I was five when the Second World War started, but memories of that time are scant. That's probably because we didn't have a television blaring in the corner and the newspapers wouldn't have been bought every day. As children, we really weren't well-informed about anything happening outside the local area. Ballylinan was the centre of our world.

I do remember my mother putting a rug over the curtains in the house to cover the light at night. She was afraid the fighter planes would see the lights in the house and drop a bomb on us. This was before electricity came in, so it's not like the house was lit up like a Christmas tree, but I suppose it was safer to take precautions like that.

There was a great fear at the time that the pilots would think they were in England, rather than Ireland, and attack us. And there was good reason for that fear. Three women from Campile, County Wexford, were killed on 26 August 1940 when the German Luftwaffe dropped bombs on the Shelburne co-op in the village. There was some debate about whether it was an accident or a warning from Hitler to keep our neutral status and not be supplying Britain with food-stuffs. Whatever the reason, in 1943 the German Government paid £9,000 (more than €11,000) to the Irish State in compensation for the co-op bombing.

In early January 1941, German bombs were dropped on several counties, including our neighbouring counties of Carlow, Kildare, Wexford and Wicklow. Three members of one family were killed when a bomb fell on the Shannon family home in Knockroe, near Borris in County Carlow. That was a bit too close for comfort for anyone living in the midlands, so my mother wasn't overreacting when she pinned the rugs across our windows. Hitler's Government said Ireland had been mistaken for the west coast of Britain. Then on 31 May of that year, two German aircraft bombed Dublin, hitting the North Strand area and killing twenty-eight people. They also dropped bombs on the Phoenix Park and the North Circular Road. West Germany later accepted responsibility for the North Strand bombings and by 1958 it had paid £327,000 (more than €415,000) to the Irish State in compensation. I'd say that some of this news filtered through to Ballylinan, but my parents probably kept it from us in case we grew worried.

There was widespread rationing at that time because imports were affected by the war going on across the water. Taoiseach Éamon de Valera declared that the country was in a state of emergency in September 1939 and he gave Seán Lemass the job of Minister for Supplies. The rationing started with tea, flour, butter, tobacco and shoes. The shopkeeper giving out the tea

9

rations was a shrewd operator. He used to make people buy some sort of table sauce before they'd get their rations and, sure, no one wanted the sauce at all. The hedge near the shop would be full of bottles of sauce that people abandoned as soon as they got their tea ration. It was rumoured that the shop-keeper would retrieve the bottles and sell them again.

I suppose everyone was trying to make a living in what-ever way they could. Lemass introduced bread rationing in 1942 and, because flour was already rationed, this had a ser-ious impact on people trying to feed large families like ours. Luckily for us, my father knew Ginger Dooley, who had a bakery in Athy. Ginger would meet Dad up the road at Killyganard Cross and give him half-a-dozen loaves, so we did better than some other families.

Despite all the rationing, my parents always managed to provide for us. That was especially the case at Christmas, when there was a tremendous effort to make sure everyone had everything they wanted to eat. There was a real party atmosphere. Sure, every day is a party now for children, but our treats were few and far between. And they were all the more memorable for it. One of my fondest memories is of waking up one Christmas morning and seeing a Little Red Riding Hood doll at the foot of my bed. I played with that doll for a very long time and I'd say I appreciated her ten times more than any youngster appreciates something they might get today. What's seldom is truly wonderful.

The Christmas pudding was a very big production in our house. We had a great tradition of everyone in the house giv-ing the pudding a stir when it was being made. The stirring was supposed to bring good luck and we'd always hope that everyone would be together again the following Christmas. To this day we would never dream of cooking the pudding without everyone giving it a stir first. My mother would make it in a calico cloth, tied at the top. She would put it on the range the night before Christmas and it would be boiling away all night. When we awoke on Christmas morning, the house would be filled with the smell of Christmas pudding. Could there be any nicer way to wake up?

We always went to midnight Mass in our parish church in Arles and that, for me, made Christmas. We travelled in the pony and trap and it really was magical. First, there was the excitement of being allowed to stay up so late. That was a novelty in itself. And we had to dress warmly because it was late at night. The moon was often very bright and there would be such silence all around once we stopped chattering. All that you'd hear was the clip-clop of our much-loved grey pony Nellie going along the road. She was a lovely stepper. It was so cold, you'd nearly see the frost hanging in the air. If it was icy, we'd go on side roads where Nellie wouldn't slip as easily as on the main road.

The church was always packed and you'd even have people standing on the altar. That's hard to believe now. After Mass everyone would gather around for a chat. No one was rushing at that hour of the night. Then when we got home, we'd be starving. Mam would take a few slices off the freshly cooked ham and it was gorgeous.

I think some of the magic was lost when Mass was moved back from midnight. I know there was good reason for it, with people coming into the church from the pub and all of that, but it's a pity to have lost something so special.

Christmas Day was one of the few times we'd see Dad really relaxing. Of course, the animals still had to be fed and looked after, but otherwise he put his feet up. On St Stephen's Day, my brothers dressed up in old coats and dresses,

disguising themselves as best they could, and they went on
'The Wren'. They called to neighbours, rattling a tin for
money to bury the wren and, if they were lucky, they might be
given something nice. But it was back to the farm work for
my father that day. We had a mixed farm of tillage and livestock and everyone did their jobs, especially on the big days
when we were drawing in the hay or doing the threshing. We
loved the threshing because we would get a few days off
school for it. The neighbours would come over to help, which
meant a lot of work in the kitchen for my mother. I remember
how she would send us out to see how many men were outside for dinner. Everyone would be fed, that was for sure.

I always liked bringing the refreshments to the field with
my mother when we were doing the hay and tillage. We had
barley, wheat and oats – the latter was grown for horse feed.
We kept horses for ploughing and I always remember the
warm, sweet smell from the horses when you'd go into the
stables. We didn't have access to flasks, so the tea would be
carried in a bottle with the tea towel around it to keep it warm.

A regular-sized apple tart wouldn't go far with the crowd
my mother had to feed, so she used to make apple tarts in the
big roasting dishes. The apples and cloves were a lovely treat.
An odd time she'd throw in a bit of mincemeat, too. The
men would nearly see you coming a mile away and they'd be
waving to speed you up. Food eaten in a hay field in the middle of summer always tasted nicer. There was nothing like it.

Then there would usually be a dance in the house when
the threshing was done. We'd have all the neighbours in and
there would be music and dancing on the cement floor in the
kitchen. That was great fun. The men would get a bottle
of stout and there'd be a sing-song. I suppose it was a way of
acknowledging all the help.

Another annual day off school was the much-loved Spring Show in May. Younger people might not remember that event. It was up in the RDS in Ballsbridge and was one of those occasions when the city and country came together. As well as the cattle shows, there were displays of farm machinery and mechanical aids to make farming life easier. The first mass public demonstration of television was held at the show by the Pye television company in 1951. I believe people queued for ages to see the newfangled device.

The show was seen as an educational day out and you can imagine how excited we'd get at the thought of heading up to Dublin for the day in the Austin A40. I was very fond of the Spring Show and was sorry to see it ending in 1992. It hurts me when people say the National Ploughing Championships killed it off. We were never in competition with it, but I think perhaps its location in Dublin made it problematic as the roads got busier. People didn't want to be driving cattle trailers and big machinery into Ballsbridge in rush-hour traffic. I don't think the RDS blamed us for the end of the show, though, because I was made an honorary member of the Royal Dublin Society in 1994. That was a wonderful honour to get and we had a lovely day out.

I always loved growing up on a farm, not that it was an easy life, but it was a relaxed life. Back in those days, we weren't taking any dictation from Brussels about European Union rules and regulations so you were, more or less, your own boss. How things have changed. Now you are told when to cut hedges and spread slurry. And if your cattle are over a certain age, you are penalized. I'd love to know what my father would have thought of all that.

We had thirty-two dairy cows and we all helped with the milking. The only equipment we had was the three-legged stool because it was all hand-milking in those days, of course. The robotic milking machines hadn't even been dreamed of in the 1940s. We'd be sent down the pasture to bring in the cows for milking in the evening. Like many farms, the fields had names. We had the Beech Tree field, and it was under that tree that the men would sit and have their food when they were at the hay. There was the Douglas, named after the river going through it, the Sheep field, the Hill field and the Long Meadow. I'm sure every farming family has a similar collection of field names. We also had the Crompán, an Irish word for low-lying land near a river. It's nice how these Irish names survived, even though we weren't a big Irish-speaking community in Ballylinan. I was good at reading Irish, but never any good at translating it. I would have loved to have been better at it. My sister Betty was good at it, while our brother JJ was a whizz at Maths. Betty used to help him with his Irish and JJ would return the favour with the sums homework.

Some of the milk was kept for churning and my youngest sister, Eileen, was great at the butter-making because she studied it when she went away to do a domestic economy course in Ardagh in County Longford. We kept hens too, and geese and turkeys. We had to help with the plucking of the turkeys and we'd have to clean them out and get them oven-ready. I never had any objection to doing it, but I'd say young people today wouldn't know where to start if they were presented with a turkey and asked to dress it. We all had our little jobs to do on the farm. My favourite job would have been bringing the tea to the men in the field because there was always great chat going on. I suppose my least favourite job would have been carting buckets of water from the pump to the house in the years before we had running water indoors.

Sick animals were my mother's department because she was great at nursing them back to good health. People back then had great old cures for animals and humans. We children often got doses of cod liver oil, which we weren't clamouring for. And if you had an earache, you could get a small bottle of olive oil in the chemist. You wouldn't have found olive oil in a supermarket back then. It was purely medicinal! My mother would put it on a spoon and heat it over the fire and slowly drop it into the sore ear. The trickle of warm oil in your ear was a strange sensation, but it seemed to do the trick.

We got electricity in 1954 and that changed everything. My God, it was a beautiful experience when the light came on for the first time. We couldn't get over the fact that you pressed a switch and the room was flooded with light. Everything was unbelievably bright. Before we got electricity, the Sacred Heart lamps, run on paraffin oil, would have thrown out a small bit of light in every room but you brought the little candles with you everywhere you went at night-time. Then the Tilley lamps came in and they were an improvement, but you'd still have to be sitting near them if you wanted to read. It was such a change to have light coming from the ceiling instead of a lamp on the table. We nearly blinded ourselves looking up at it.

There was great excitement when the men came to connect the house to the grid. The floorboards had to be lifted to bring in the wires. They were all put back in place afterwards, but there was one floorboard that never fitted back properly. Where do you think that one was? Right outside Mam and Dad's bedroom, of course. Every time we walked on it, it creaked, so we had to be very careful to skirt around it if we arrived home very late at night. Dad would often wake up and say to Mam: 'Are they home yet?' Keeping the peace, she'd say: 'Sure they came in long ago, James. You were fast asleep.' It's a good job he never got up to check on us because he wouldn't have found us in the beds!

My father was great for moving with the times once the electricity arrived. As soon as he could afford it, he bought a television and gradually we got all the modern conveniences: the washing machine, the fridge, the cooker and the vacuum cleaner. The arrival of each item was greeted with great excitement. I think Dad was the first in the locality to buy an electric welder. And he was definitely one of the first to get a combine-harvester. My brother JJ collected it and drove it home from Cork. It was a Lanz combine and people came out of their homes to see it, such was the novelty of the new machine.

Our home was fairly typical of the time. The big old range, throwing out heat day and night. The dresser with the willow pattern plates and cups on hooks. The long kitchen table with the long forms, or benches, on either side and the drawers on both ends. One drawer was for cutlery and the other was for odds and ends. Dad kept all his songbooks in the second drawer. He loved a sing-song, even though he didn't have a note in his head. He wouldn't let us leave the house without singing for him. Every time he went to Dublin he went to Walton's music shop and bought a songbook. My father loved a good ballad. Count John McCormack was one of his favourite singers. While he couldn't sing in tune, he still knew all the words. If you were singing and forgot a line, you could be sure he'd fill it in.

Now, Mam was musical. She played the melodeon and *When Irish Eyes are Smiling* is one tune that will always remind me of her. *The Old House* and *The Spinning Wheel* were more of her favourites. She also played the violin. I could only play *Pop Goes the Weasel* on it. When it came to the melodeon, I could only do the start of *When Irish Eyes are Smiling*. I could never get further than the first four lines.

Perhaps Mam picked up her love of music from school. She was educated with the Brigidine Sisters in Tullow, County Carlow. I recall hearing that she was very good at basketball, despite her small stature.

It was most reassuring to come home from school every day and find Mam in the kitchen. It's so different for children now, but I remember being surprised if Mam was not in the kitchen. We nearly expected her to go nowhere else in the house. If she wasn't there, we'd be calling all over the house for her. But at the same time, we knew that if we opened the bottom oven in the range, our dinner would be there, waiting for us. We cured a lot of bacon – I suppose it was the cheapest meat – so bacon and cabbage was a staple dinner in our house. Colcannon, too. And there were a lot of stews. There was nothing as good as coming home from school on a cold day and smelling the hot stew in the kitchen.

Money was scarce for nearly everyone in those days, but we were never short of the real necessities. My father's mantra was always: 'As long as everyone is well inside the house, everything else will take care of itself.' Now in those days it would be a big setback if an animal died, but he took the view that it would be far worse if there was a sick child in the house. You could always replace an animal. He would be terribly upset if someone wasn't well, but that didn't happen too often. They rarely had us with the doctor.

Dad had experience of sudden death at home because his own father died when he was just eight years old. He used to drive the steam engines for threshing and apparently he was coming down the road and one of the wheels slipped into a bit of a dyke. My grandfather was a fine, big, strong man and he got out to edge the wheel out of the dyke. He came home that evening not feeling well and he died the next morning. They said he had burst his heart with the effort of lifting the wheel.

It was sad for Dad, being only eight years old and having no brothers or sisters. He was terribly lonely. Later on, he was sent to Castleknock College in Dublin but hated it, and ended up in Newbridge College in County Kildare. His mother was an extremely genteel person, originally from Ballymore Eustace, and it must have been a huge change for her to be living in Clonpierce with no family around.

I was really very close to Mam. We were her whole world and I never once heard her looking to go anywhere. It wasn't the done thing for women to go out much, or go away for weekends. She was a terribly quiet woman, very loving and great fun. I remember the boys lifting her up and carrying her around the kitchen and she'd be laughing and saying: 'Put me down, you'll hurt me.' I don't believe she ever said a cross word to us, and that's some feat when you remember there were eight children in the house. Can you imagine the noise we'd be making?

I remember she'd say: 'Your father's coming. Now you'll stay quiet.' And we would. We never back-answered our parents. We'd never even dream of it. But if Dad went to chastize us, Mam would intervene. It wasn't that Dad was cross, but he was the controlling power in the house. I remember asking him if we could go to a dance when we got older, and the lecture we'd get! Poor mother wouldn't say a thing to us. But having said that, he would give us the last shilling he had, he was so generous. He would go short to help us. He loved the company of a big family, being an only child. I remember him saying how he often used to stand at his own gateway when he was a child, waiting for someone to come and visit his parents.

Thanks to Mam and Dad, we never had that problem. We always went to school together on the pony and trap, driven by a man called Mike Bolger who worked with the horses on the farm. If the evening was fine and they were busy on the farm, we'd walk home. There would be ten or twelve of us and I have such happy memories of those times. We would dawdle along without a care in the world, us, the Walshes, the Behans, the Julians and the Donoghues. We're still very good friends to this day. There's something about the friends you make at that age. You never forget them. It could take us an hour to get home. Sometimes we'd take off our shoes and walk in our bare feet for comfort. Of course, we weren't saints and we got up to a bit of devilment every now and then. There was an orchard on the way home and while we had our own orchard, the apples from the other orchard always tasted sweeter.

There was the time we got cigarettes somewhere and decided to try them. Sure, we nearly killed ourselves with the smoke. Then we decided to chew ivy leaves to get rid of the smell off our breaths. How we didn't poison ourselves, I'll never know. I'm not sure if it was the taste of the cigarettes or the taste of the ivy, but I never took up smoking.

We got up to a few antics, that's for sure. One day, Stannie and Oliver decided they would bring the donkey upstairs. It wasn't clear what their plans were once they had negotiated the stairs with her, and I don't think they'd considered the logistics involved in getting her back down again. Anyway, they opened the front door, got her up the front steps and into the hall. They were trying to coax her up the stairs when Mam discovered what was going on and put a stop to it.

The five boys shared a very big bedroom, so you can imagine the devilment that went on there. The personalities were so different. I won't name any names, but we always knew who was causing the trouble and who was trying to calm it down.

Myself, Betty and Grandmother Annie shared another room, and Eileen was in my parents' room in her early years. There was a beautiful marble fireplace in our bedroom and Mam would light a fire in it during the winter months. We loved to sit around it and we spent many a night telling tales and chatting about the goings-on around the neighbourhood. No stone was left unturned during our fireside chats, whether we were talking about who played well at a match, who got in trouble in school, what was happening on the farm, or when was the next social event in the parish.

I didn't have one best friend among my siblings – it really depended on what was going on at the time. I used to go to dances with Betty because Eileen was that bit younger, but then I had a lot in common with Oliver because of the GAA. Stannie was my mechanic, Paddy always made his Volkswagen available to me before I got my own car, and Gerald was the quiet one who would give us a hand with homework. JJ was an early riser, like myself, so we often fell into step together in our daily activities. And sure Eileen and myself played camogie together, so we were always going off training together and preparing for matches. Eileen was very fond of sewing and I enjoyed rug-making and embroidery. I even made curtains for the home place at one stage.

We were a very happy family, very united, and we still would be like that to this day. When I think of my younger days, I remember the great sense of togetherness. There was nothing nicer than sitting around the fire at night, often with someone sitting on the arm of your chair because there weren't enough seats for everyone.

We loved to play cards. Our favourite game was '25', and Eileen would claim that I was the greatest thief going when it came to cards. I would be obliged to disagree, of course. My mother would often say it was a good job we lived a long way in from the road. She was relieved no one could hear the shouting when the card games got going. And when things got fairly heated, she would threaten to throw the cards in the fire.

On summer Sundays, when we were in our late teens, a group of neighbours and cousins would come together and we'd pick a spot for a picnic in the neighbouring townland of Gurteen, or we would go to Kilmoroney estate, which was the Georgian home of the Weldon family, now in ruins. Colonel Sir Anthony Weldon was the British commander of the troops in Limerick during the Easter Rising of 1916.

Dad often brought us in the pony and trap to visit our cousins in Naas – the Kehoes and Cullens of Elverstown, and the Brennans of Delamaine. Obviously we could not all fit in the trap, but the girls seemed to get priority on that trip. If it was cold on the way home, the youngest member of the family would be put on the floor of the trap and covered with a blanket. Nellie the pony never faltered or let us down.

The rosary was said every night in the kitchen, or in the bedroom where the fire was lit during winter. Sometimes we'd be wishing Mam would forget about it because we wanted to go somewhere. We'd kneel down with our elbows resting on the kitchen chairs and you'd be hoping you'd get through your mystery of the rosary without breaking out laughing because someone was pinching you or trick-acting. Our intentions were good, though. We weren't unique in our devotion to the rosary. Most families I knew said the rosary every night. In fact, there was a letter-writing campaign to the *RTÉ Guide* – then called the *RTV Guide* – in the 1960s, asking Teilifís Éireann to clear their schedule at 7.00 pm to allow families to say the rosary. Damian Corless, in his book on 1960s Ireland, *Hopscotch and Queenie-I-O*, recalled how parents wrote letters to the magazine explaining how hard it was to say prayers when their children were afraid they'd miss something good on the television. They suggested that ten or fifteen minutes of television should be devoted to the rosary every evening. Who'd have thought we'd arrive at the day when people were complaining about the one-minute Angelus being broadcast?

My father loved the Angelus, and to this day I think of him when I hear the Angelus bell. He would always stop whatever he was doing when the Angelus bell struck, and bow his head. Religion was a big thing in our house and in our school. We were always asked on a Monday morning if we were at Mass the day before. There was no question of not going to Mass, unless something extraordinary had happened or you were very sick. There's a part in the poem *The Planter's Daughter*, by Austin Clarke, when he's talking about how everyone admired the planter's daughter, and he describes her as being the Sunday in every week. That gives you a good idea of the importance of Sunday in our childhood. It was the pinnacle of the week. We had the nicest dinner on Sundays. Our best outfits were always kept for Mass on Sundays and taken off as soon as we got home. And if we got something new, we looked forward to wearing it to Mass first, as a mark of respect. You were also guaranteed that everyone would see it because everyone else was at Mass, too.

Sundays have changed for everyone since then, of course, and it's a working day for many people now. People are flocking to the shopping centres instead of the churches and they are saving their new outfits for that. But I do often think with longing of those Sundays when everyone downed tools, came together as a family, and took it easy.

We still try to keep Sunday as a family day and only really work on a Sunday when the Ploughing is looming or wet weather is coming in and crops have to be saved. I am a strong believer in the family sitting together for meals, and it would be most unusual for me to dine alone. We are still a great family to get together for all celebrations, from anniversary Masses to Communions and Confirmations, and I'm at my happiest when I'm in the middle of it.

I often reflect on how lucky I am to have had the childhood I had, and it has coloured everything else in my life since. I'm very grateful to my parents for giving us that sense of unity and belonging. I feel it has moulded me as an adult, and I hope I've managed to pass on that spirit of family unity to the generations coming after me.

2. Camogie and carefree days

You might not think it now, but I was a total tomboy as a child. We were outside all the time and I suppose I had to be hardy, being the fifth of eight children and surrounded by boys. One of the benefits of coming from a big family was that there was always someone around to play with. On top of that, we had our first cousins next-door, so we'd never run out of playmates. Groups of lads would come over to our house for football matches and we'd play until we couldn't see the ball in the dark. I was often put in the goals because I was younger. Hopping the handball off the side of the house was another big pastime of ours.

When we'd get away from the boys, we girls loved playing Ring-a-rosie and we'd spend hours making daisy chains down in the grass. The length of them! And we'd nearly cry when they'd break.

Then there were the cricket games. I don't know how we ended up playing cricket, as I'm sure it wasn't a common game on farms in the 1930s and 1940s. Although, a cricket club was formed in Athy in 1872 and was said to be one of the first cricket clubs in the county. And of course, Kilkenny had a long history of cricket and Lory Meagher's father, Henry, was a well-known cricketer with Tullaroan in the late 1800s. Anyway, my brothers fashioned a cricket bat and we used a flagstone in the place of a wicket. We'd spend hours mastering the game and we couldn't have been happier. That early experience with the cricket bat might have led to my great love of camogie and might explain why I went on to play for Laois and Leinster.

After I left Mrs Fleckney's knee, I negotiated my way through national school as best I could, dodging the ruler whenever possible. Then I headed for the bustling town of Athy for the next stage in my education. I cycled the four-mile journey to St Brigid's Technical School on my High Nelly. Every morning I'd be looking out the window and praying there would be no rain because there was nothing worse than sitting in soaking wet clothes for the day.

The school had a camogie team and it immediately caught my interest. All those days running around with the makeshift cricket bat began to pay off and I loved the sport. Laois has had a funny relationship with camogie. It has been played for over 100 years in the county, but it goes from being extremely popular to the point where you'd struggle to get a team together.

There was no club in Ballylinan when I started playing at school. So, when I left St Brigid's I was faced with a choice: give up camogie for good, or start a club. If you know anything about me, you'll have a fair idea about what I did. I started agitating to get a camogie club going in Ballylinan. Now there was absolutely no interest in hurling on our side of the county at that time, so it was a bit of a challenge. But I succeeded in rounding up twenty girls who were interested and we invited the county board to meet us.

The ball – or the *sliotar* – was rolling and we formed a club. We started training two evenings a week, and you wouldn't believe how much I looked forward to those sessions. Michael Dempsey trained us, and if that name sounds familiar it's because his nephew, also Michael Dempsey, plays a key role in the management team of the Kilkenny hurlers. I don't know how the Kilkenny lads would manage with our facilities because they were non-existent. We didn't have dressing rooms – we'd be changing behind the ditch or with the doors of cars open, trying to keep ourselves as respectable as we could.

I clearly remember the first championship match we played. I'd love to be able to tell you that this fledgling team of brave young women thrashed the seasoned opposition, but I'm afraid we were beaten off the pitch. This pattern would continue for a few years because we just didn't have the experience. But then we set up two clubs nearby so we could have more practice games and things started to come together. We didn't have cars at the time, so we had to hire a car in Ballylinan to travel to matches. I thought it would be a good idea to get local businesses to sponsor our travelling expenses and that worked well for us.

Of course, we hadn't the money for a uniform, so we had to ask our parents to contribute to that cost. Our playing gear consisted of sky-blue gym frocks, white shirts, black stockings and black knickers that went down to the top of the stockings. And I remember this contrary old biddy of a referee calling me aside at half-time during one match to tell me that there was a division between my stockings and my knickers. This meant you could see a bit of my leg, and that wasn't allowed. I can tell you, wondering if an inch of leg was showing between my knickers and my stockings was the last thing on my mind when I was racing up the field for the *sliotar*.

I played centre-field while my sister Eileen played centre-back and those were grand days indeed. We even had a boy playing with us on one occasion. We were having a match at the carnival in Athy and we were short of a player, so we cajoled this lad, who shall remain nameless, into play-ing. He put on a gym frock and off he went. We called him Jenny. It was discovered afterwards, but it was just a bit of fun. Nothing hinged on it.

My brothers were playing hurling and football in those years, but the camogie was generating a lot of local inter-est. As we improved, my mother and father got caught up in the excitement and never missed a match. We won the county championship five years in a row in the 1950s. Camogie was flying high at this stage and we had lots of girls joining up. Dublin was the team to be feared, and for good reason. With the exception of 1956, they won every All-Ireland Senior Championship title from 1948 to 1966. That's eighteen out of nineteen titles. Antrim broke their run in 1956 and ended their run in 1967. Wasn't that some record for Dublin? And worse for us in Laois, they absolutely dominated the Leinster Championship between the mid-1930s and the late 1960s.

I remember coming up against them in a Leinster final in 1961 and being afraid of our lives meeting them. Neverthe-less, we held them until half-time, but they beat us by two points in the end. I sat down on the hurl and cried a few bit-ter tears. I wasn't the first, nor the last, player to do that. Especially where a Dublin team is concerned.

To be selected to play for Leinster was the icing on the cake for me. Because Dublin were always winning the All-Ireland Camogie Final, they would get six or seven places on the Leinster team. A county like Laois was lucky to get one place on the team, so I was thrilled to be picked. I was selected in 1957, 1958, 1959, 1960 and 1961, and we won the final in the first four years, beating Ulster in 1957, 1958 and 1959 and Munster in 1960. It was a Munster v. Connacht final in 1961, the last year I was on the Leinster panel. The only

downside of being selected for Leinster was that I'd be singled out for good marking at club games because they knew
that I was a Leinster player.

The Dublin girls were great to play with and they were
very entertaining company off the pitch. They were terribly
mischievous. I remember when we were going up on the
train to play Ulster in Casement Park in Belfast in 1959 and
the ladies started singing *Kevin Barry*. I was afraid of my living life that we'd be blown up, but the more concerned I
grew, the worse they got. We had a wonderful time travelling
the country, playing in places like Casement Park, Pearse
Stadium and Parnell Park. Those were such happy days,
when we hadn't a care or a worry in the world.

Despite spending a few nights a week training, I still
found time for a few outings to country dances. We would
cycle to the halls, so there wasn't too much finery involved.
There was one dance hall in Timahoe, County Laois, that
was practically a hut. And we often cycled to halls in Lug-
gacurren and Ballyshannon – the Kildare one, not the
Donegal one. I wasn't going that far for a dance. They were
mostly local bands, but I remember Brendan Shine came to
Timahoe one night and, God, we thought it was so fabulous
to have someone like him in Timahoe. There was a whiff of
glamour and showbiz off him. We also went to the Ritz Ballroom in Carlow and Dreamland in Athy. There was more
glamour and dressing up involved in those events.

Alcohol was never a feature of my nights out. I'm very
proud of the fact that I haven't broken my Confirmation
pledge yet. Nor am I likely to take up drinking at this stage
in my life. In fact, five out of the eight of us in the family
decided we wouldn't drink. Now, I have nothing against
alcohol and I'd socialize with people and buy them a drink,

but I never felt the desire to take a drink. So long as alcohol doesn't interfere with family life, I don't mind people having a drink.

When we were very young, my father liked to take a drink. I was too young to remember but Betty, who was the oldest, told us that she had asked Dad not to drink because she didn't think it was a good example to the children. He was a quiet man with drink and Nellie the pony knew her way in and out of Carlow, so he always got home safely. Now he wasn't an alcoholic by any stretch of the imagination, but he mustn't have liked the effect that drink was having on him because one day, of his own accord, he suddenly went into the nuns in Athy and took the pledge. That was it. He never drank again. I really admire what he did for us. His whole life revolved around his family and looking after us all and I suppose he could see he wasn't doing the best for his family when he took a drink. You'd have to applaud him for taking a stand like that and being strong enough to stick it out.

One by one, we learned how to drive and there would always be great competition to see who would get the family car at night. Getting a loan of the car was a big thing. I remember Dad coming out the morning after one of us had the car and walking around it to make sure there were no scratches on it.

He taught me how to drive, but it didn't always go so well. I think he was more concerned about the well being of the car than about me. He wouldn't be the first father to think like that. I was as nervous as a bag of cats when I started driving and, to be fair to him, he was quite patient. He took me out into a field, away from anything I could hit. He kept me there for a long time before he judged that I was competent enough to meet other vehicles on the road.

My beloved first car was a black Morris Minor, registration ZN 1437, purchased in 1961. I can't remember how much it cost, but I know that it was heavily subsidized by my father. He was always very generous if wc needed anything and he loved to be able to help us out. I was probably one of the first in our camogie team to have a car, which was very important when it came to the away matches. I thought I was the bees' knees and I loved carrying girls to camogie matches and the like. There's not a whole lot of room in a Morris Minor, so they'd be hanging out the doors and the windows. I remember being so proud to be able to bring Mam away for little holidays when I got the car. One time I surprised her by taking her away to Arklow for a weekend. I had to send word back to Dad to tell him where we were, in case he thought she'd been kidnapped.

When the time came to trade it in, my brother Stannie brought it into Naas to meet the dealer. To Stannie's disgust, he offered him £10 for my car. £10! And Stannie was so insulted on my behalf that he said to him, 'Sure it's not a wheelbarrow you're buying' and he turned the car and brought it home again.

I sold it afterwards, but I was terribly sorry I ever let it go because it would be an antique now. I loved that old car. I traded up to a Ford Anglia after that, but nothing brings the same thrill as getting your first car.

The whistle was blown on my camogie career in the early 1960s. I never gave up camogie. It gave me up in the end. I would have played forever, but you just know when your time is up. It gave me some of the happiest years of my life and I'll always look back on them with pleasure. After I hung up my hurl, the scene totally changed in Laois and ladies' football took over from camogie entirely.

To mark the GAA centenary in 1984, we were all asked back to Ballylinan for a match with the younger players. Lord, we had such fun that day. We marched down the street behind the band and it was great sport to get back on the pitch. My daughter, Anna Marie, played with the younger players on the opposition team that day and the young ones said we were absolute butchers. I would dispute that!

But the challenge game left me under no illusions about the decline in my skills. I'd be swiping at a ball and not connecting with it and saying: 'What has happened to me at all? I've lost my knack.'

As I was involved in getting camogie going in Ballylinan, I found myself taking charge of the club. I was also county secretary for a while and I was elected vice-chairwoman of the Leinster Camogie Council. Unknown to me, I was learning skills at meetings that would stand to me in the following years.

I love sport and I really admire the work that the Gaelic Athletic Association does. The training and the commitment that they give to young children is absolutely wonderful. And it's all voluntary. I love that community spirit. And I love watching the matches, even though it's not good for the stress levels sometimes. If anyone came into the house during the All-Ireland Hurling Final between Tipperary and Kilkenny last September, they would have very quickly reversed back out the drive and said: 'That woman is going stone mad.' I have the height of admiration for Kilkenny, but I was cheering for Tipperary because you like to see someone else getting a chance now and then. I was shouting at the television and watching the clock like a hawk. They always say you've never beaten Kilkenny until they're actually on their way home because the game can change so fast in the last few minutes. The win for Tipperary, by nine points in the end, was a good result for the game and I was hoarse for the night after it.

While I was travelling the country playing camogie, my brothers and sisters were starting to go steady and get married. Every one of us settled down within a 12-mile radius of the home place. It was probably very unusual for the early 1960s that not one of the eight siblings needed to emigrate to find work. We were very lucky, and we have my father to thank for a lot of that. He had great foresight and he bought a farm for four of the five boys. Gerald's wife, Marie, had no brothers and she had inherited land, so Dad didn't need to look after him.

There was never any suggestion that he should buy a farm for the three girls because it just wasn't the done thing at that time. You gave daughters what we called a fortune, which I suppose was a kind of a dowry. Your father would pay for the wedding breakfast and give you a generous gift to help you start your married life. We certainly would never have expected him to buy us land.

Looking back now, it was some achievement for my father to do that. He might not have the price of the land at the time he was buying it, but he'd manage. I remember when he bought a farm for Stannie and Mam said: 'How will we finance it, James?' And he said: 'Lizzie, if I can keep the lads working together, the farm will pay for itself.' And he was right.

My father would have been proud of them because every one of them continued farming. But they all did it in a different way. Paddy was a great cattle man, an out-and-out farmer. JJ, the oldest brother, was an exceptional farmer so he took over the home place. He could turn his hand to anything and could build the finest stone wall. He was always improving things and he had a great head for figures.

Gerald was Grandmother's favourite and we always teased him, saying that Grandmother spoiled him. We used to say he hated getting his hands dirty. He ended up working with Glanbia in the financial area, but he also farmed.

Stannie was great with machinery, but he wouldn't look at a cow. He's still farming away, mainly sheep and tillage. Oliver always looked after the cows and calves when we were young and he still does that to this day, on his own farm.

The oldest child, Betty, became a nurse and she ended her working life as a sister in St Vincent's Hospital in Athy. She wouldn't do office work, not if you gave her €1,000 a day, but she was a brilliant nurse. She would often come into the office to visit me and look around and say: 'In the name of God, how do you stick this kind of work?' She married a farmer, Ned Whelan, and she always loved her cattle and milked cows for many years.

My youngest sister, Eileen, is well known through her work with the Irish Draught Breeders' Association and it amuses me when I go to the Dublin Horse Show and she introduces me as her sister. It's a nice change to be in the background and to see Eileen holding court and being the centre of attention. I think she gets a kick out of it too.

After Eileen gave up camogie she took up hockey and was a very fine player indeed. When I put away the hurl, I was encouraged by friends to take up golf. I tried it for a bit, and while I had a good long stroke on the ball, I just didn't have the patience for the game. I found golf to be a very slow game after the excitement of camogie. My nearest nine-hole golf course was Athy – it's now an 18-hole – but ladies' day was Thursday and this did not suit me because it was always a busy day at work and golf is very time-consuming. After about a year I gave it up. I found it hard to accept the quietness of golf. When you score a point or a goal in camogie there's great excitement and delight, but the celebrations were very restrained in golf.

So that exit from golf marked the end of my sporting career. I've been so busy with work that I haven't had time to take up anything else. I suppose all that running around is enough exercise for me these days. But I love to look back on my camogie days and I'm very proud to have played for my county and province.

3. Ploughing my own furrow

Even from a very young age, I loved a bit of organizing. I was always parcelling up things, thinking I would be a shop assistant one day. When we were young, Dad made us a babby house – I suppose you'd call it a playhouse now. It was in the shape of a beehive and we played in it all the time. We used pebbles for money and played shop. That's when I practised, and perfected, my parcelling-up skills.

There was one moment early on when I realized that I was good at organizing. Muintir na Tíre was very strong in our area at the time, and it won't surprise you to hear that I decided to join up. Muintir na Tíre (which is Irish for 'people of the country') is a community organization that was set up in 1937 by Canon John Hayes. He was trying to encourage a sense of community spirit and neighbourliness among rural dwellers and to get people to think about ways in which they could help themselves. Now, you have to remember that Ireland was a very rural country at this time. When I was twelve, in 1946, more than 60 per cent of people lived in the countryside or in small towns and villages. We had very few diversions, so organizations like Muintir na Tíre were embraced because they gave us somewhere to go and something to do in the evenings. There was a junior section in Ballylinan and it was the first thing I got involved with as a young teenager.

We had to organize various things, like functions and quizzes and little get-togethers. When I was about fourteen I was put in charge of giving out the oranges at the children's Christmas party. Getting an orange was a big treat and an exotic novelty in those days. I was given this timber crate of oranges and I duly went ahead with my task and gave them out. Everything went smoothly, but then the chairman, a retired commandant, came up to me. He was a very serious man but, to be quite honest, I don't remember feeling intimidated by him. He told me I had given out the oranges too early, and I said to him: 'Thank you very much, but I'm in charge of this section and this is my decision.' I felt that if someone gave you the responsibility to do something, then they should trust in you to do it right. I liked the feeling of organizing something and doing it right. He had the height of respect for me afterwards. I suppose he thought, Well this young lady won't be pushed around.

While I knew that I liked organizing things, I had no idea what I wanted to do after school. Very few people did the Leaving Certificate in those days and many left school much earlier than that. According to the Central Statistics Office, 35,000 children aged fifteen or younger were gainfully occupied in 1946, and half of these were working on the farm. In 1950, when I turned sixteen, only 4,500 children sat the Leaving Certificate. This was back in the days when secondary schools were fee-paying and only a minority could afford to educate all their children there. University was only for the elite, the children of doctors and judges. In 1950, the entire university population of this country was just 7,900. The student population of my nearest third-level college, Carlow Institute of Technology, would be almost that now.

Most of my sisters and brothers went to secondary school but I went down the vocational route, which was funded by the State. I suppose it was felt I was a more practical type of person and those skills would suit me better. I spent two years doing a secretarial course in St Brigid's Technical School in Athy, learning things like shorthand, typing and book-keeping, instead of doing the Leaving Cert. I know it's not widely taught now, but I found shorthand extremely useful over the years. It was great for taking minutes of meetings. I still use a few of the outlines from the *Gregg Shorthand Dictionary*, but you forget a lot of it when you don't use it.

Typing was my favourite subject. I worked up to a speed of about 120 words per minute. We had a row of typewriters in the classroom and there was always a race to get to them in the morning because there were a few machines that should have been put out of their misery years earlier. The townies, as we called the people from Athy town, would arrive early because they didn't have to travel four miles on an old bicycle, so they would have laid claim to the best typewriters by the time we arrived. The country children would be left with what I called the grubbers — that's the word we give to a heavy implement that you till land with.

I had my first office experience during one summer when I worked in Malcomson Law solicitors in Athy. I must have been sixteen at the time. I still remember being in this dark little back room, surrounded by piles and piles of files everywhere, from the floor upwards. They were very rigorous about grammar and punctuation when you were typing up letters, and woe betide if you put a comma or fullstop in the wrong place. It was an office experience, that's for sure, but I knew it wasn't the future I wanted.

It's funny how small, seemingly inconsequential things can shape our whole lives, but that was definitely the case with me. I finished the secretarial course in 1951 and didn't have a clue what I should do with my life. I was seventeen years old. I was a home lover, with no intention of flying the nest, so what happened next was fortuitous. At that very time, a man needed some help in his office. A month later or earlier, and he may have found someone else and God knows what direction my life would have taken.

That man was John James Bergin, better known to everyone as JJ Bergin. He was co-founder, with Denis Allen, of the National Ploughing Association (NPA), which was in existence just twenty years at that time. His biggest job was staging the National Ploughing Championships once a year. Without him, I would have no story to tell. JJ and my father were friendly, and JJ asked him if 'the young lady', meaning me, would give him a hand in the office. Dad would have been disappointed if I didn't do it and I didn't want to embarrass him, so I said I'd go down. I set off for JJ's house, in Maybrook, Athy, on a Friday morning thinking: 'Well, it's just for a few days.'

My lasting memory of meeting JJ was of seeing this old man with a big, grey head and a moustache. I had no idea of all he had accomplished. He had so many strings to his bow. After his death, an article in the *Carlow Nationalist* listed his achievements and I hadn't even heard of some of them. As a progressive tillage farmer, he had published helpful pamphlets for farmers and even edited a magazine called *The Farmers' Guide*. During the Second World War, he was a great proponent for the use of a wheat-maize mixture for bread, which helped people at a time of strict rationing.

He was a civil engineer but had also worked as a mining engineer before the First World War. According to the article, he invented several types of agricultural implements and many of them were copied and improved upon by foreign manufacturers. Apparently, he was a great actor and was one of the leading members of the Athy Hibernian Players stage group. He was also very active in Muintir na Tíre. In his youth, he was one of the best-known track cyclists in the country and it appears that cycling ran in the family. Research by Athy's local historian, solicitor Frank Taaffe, found that JJ's older brother, Andy, had ridden one of the first Penny Farthing bicycles around the town. In later years, JJ organized athletics competitions for local teenagers. He also set up a Pipers' Band in Athy and was very active in the Ancient Order of Hibernians. During his research, Frank Taaffe found that JJ had run, unsuccessfully, for the Dáil twice – as a Farmers' Union candidate in 1922 and as an Independent Farmers' candidate in 1927. Then he struck lucky in 1928 by winning a county council seat for Fianna Fáil. As Frank says, it would take a full book to do JJ justice.

I knew none of this, of course, and I came home and said to Dad: 'Oh, he's a very elderly man. I'm not going to work there.' JJ was only in his early seventies at the time. Now that I'm eighty-three, someone in their seventies is just a whipper-snapper, but when you're seventeen you think everyone is old. Anyway, for some unknown reason I went back on Monday morning. I was in the middle of some work, so I said I'd stay on a few days longer to finish it. Then a week went by, and a month, and the rest is history. Some sixty-six years later, I still haven't left. My meeting with JJ was the start of a wonderful friendship with the Bergin family. The office was in their home and I quickly became part of the family.

JJ was a very nice man to deal with and he made work easy. His wife, Kathleen, and their six children were lovely people and a strong bond developed between us all. Their son, Ivan, was very involved with the National Farmers'

Association, later to become the Irish Farmers' Association (IFA). Of course, the National Farmers' Association grew out of Macra na Feirme, the young farmers' organization. It had its headquarters in Athy at that time. So you could say Athy was the birthplace of the National Ploughing Association, Macra na Feirme and the National Farmers' Association. And Ernest Shackleton, the Antarctic explorer, was born a few miles outside the town. There must be something in the water in Athy to encourage such a strong 'get up and go' mentality among its people. Ivan had been involved with Macra na Feirme before the National Farmers' Association was formed, and in the late 1940s he designed the logo that is still used by the organization today. His father must have been very proud of it as it depicts a man ploughing. Ivan was also secretary of Kildare County Show and so I fell into that work too, as the years went on.

I am often amazed at the foresight of our forefathers, particularly when I think of the mission statement that JJ and Denis set out for the NPA. It is as relevant today as it was almost ninety years ago. It states that the mission of the NPA is: 'To bring the message of good ploughing to all parts of the country and to provide a pleasant and friendly place to meet and do business.' We still strive today to maintain that ethos, and we always will.

I have to say, I knew very little about the intricacies of ploughing when I started working with JJ, other than hearing my father and brothers talking about it. Now would be a good time to explain what ploughing involves, as we sometimes forget that not everyone is as enthusiastic about the art of ploughing as we are.

Ploughing is the first stage in the cultivation of the soil. The plough slices through the earth, turning over the upper layer of the soil in preparation for the sowing of crops. When you turn over the sod, you are bringing the rich nutrients to the top of the soil, and burying weeds and any remains of the previous crop. Every farmer wants a fine, firm, weed-free seed bed, to give his crop the best possible chance to thrive, whether he is planning grain, grass seed or vegetables. And so, the better the ploughing, the more economical the crop growth is.

I suppose competitive ploughing came about because of this drive and passion for excellence in ploughing. The National Ploughing Championships was still a small operation when I joined the workforce. JJ did most of the office work, but he had a lot of voluntary support when it came to the national event. In my new position, it was my job to get JJ ready for meetings, sort out his briefcase and make sure everything was in order. One of my first jobs was to list every letter we sent out, in the postage book I also had to fold circulars – they would contain rules for the various competitions and information on the venues. All the competitors wore calico bibs with their numbers on them, so making those calico numbers was another job for me.

I progressed on to writing letters after a while. If I misspelled a word, JJ would make me write the word ten times in a copybook. That perfected my spelling, without a doubt. I'm only sorry that I can't lay my hands on that copybook because I'd love to go through it now. I wasn't a bit offended by JJ's approach, although I know it probably sounds a bit unusual. I couldn't imagine a boss doing that now, but JJ used to fill in as a substitute teacher in Athy so I suppose it was the teacher coming out in him. But he was an encouraging boss and he obviously saw I was capable of doing bigger things. Soon I was going to meetings and playing a bigger role in the running of the Association, and I loved every minute of it.

My first wage was 22 shillings and six pence a week. I thought I was so well-off. I remember sending five shillings to the Poor Clares out of that first wage packet. We had Poor Clares in Graiguecullen, on the outskirts of Carlow town, but for some reason I overlooked them completely and posted it to the order in Hawarden in Wales.

The old High Nelly was ready for the scrap heap after I finished school, but now that I was flush with cash I was able to buy a beautiful, pale blue New Yorker bicycle. It was a sports bicycle, very flashy, with a dynamo, and I couldn't have been prouder. It was the BMW of bicycles. I was polishing it from morning to night.

JJ also ran his civil engineering business from that office so I wasn't just doing work for the National Ploughing Association. The office equipment that was available to me then wasn't exactly state-of-the-art. In fact, equipment is too grand a term for it. The first typewriter bought in the NPA was a second-hand one, purchased in the late 1940s for £35. I have every reason to believe it was the one I used when I joined the office in 1951. It was a heavy, black Remington. I later got an Underwood typewriter, which felt like the height of sophistication after battering the keys of the Remington into submission.

That was the extent of our office equipment. You can imagine the lengths we had to go to when we needed to post circulars to our members around the country. I had to get a stencil, put it into the typewriter and type the circular. Then I fitted that on to the Gestetner machine, filled it with ink and rolled it to make a copy.

When the fax machine came in we thought things couldn't

get any more advanced. The idea of putting a letter in a machine and someone the other end of the country getting it seconds later was a revolution. If someone had told us what was on the way, with mobile phones, e-mail and Skype, we'd have thought they were away with the fairies. I've used them all, mobiles, e-mails and Skype, and while they have their uses, I still think there's nothing like talking to someone face-to-face.

But back to the National Ploughing Championships, and I should tell you how it all started. I bet not many of the 283,000 people who come to the Championships know that the whole thing started because of a simple wager. JJ Bergin and his good friend, Denis Allen, a Wexford man, were having a debate about the best ploughmen in Ireland. Women were not even considered at that stage, of course, but that would later change. Both men were progressive farmers and, needless to say, they both thought their own counties had the best exponents of the art of ploughing. Now, remember this was the 1930s, so they were not discussing tractors and ploughs. It was all horse ploughing then, and being a good ploughman was a much sought-after skill.

Horse ploughing is more intense than ploughing with the tractor because there has to be a great communication and understanding between the horse and the ploughman. Nowadays, someone leads the horses, but back then it was just one man and the horse, and the ploughman had to be very specific with his instructions. Even a slight delay in the horse's reaction to stop or go forward could change the whole appearance of the ploughed ground.

I remember the old story where a competitor from Northern Ireland was ploughing in our National Championships one year and he decided to borrow horses from one of our ploughmen instead of bringing his own. A problem arose very early into the competition. The horses just could not understand the Northern accent of the ploughman, and the poor old things did not know whether to start or stop. Eventually the ploughman had to pull out because he was getting nowhere.

It really is a pleasure to stand and watch the way the horse can walk in the furrow (the cut in the ground made by the plough) without ever walking on or damaging the ploughed ground. They tread so lightly and carefully, totally aware that there is a purpose to what they are doing. It's a slow process, and turning the plough team at each end of the plot has to be done seamlessly, with no distraction from the supporting crowd, especially at the National Championships.

The high level of skill involved in horse ploughing has meant there has always been lots of rivalry between ploughmen, and challenges were held as far back as the early 1800s. We've heard of one in Camolin Park, Wexford, on 20 October 1816. On that occasion, a special prize of £5 was put up for the carpenter or plough-maker who produced the best and cheapest plough made by himself and who contracted to supply the public with similar ploughs at the same price.

In 1854, the *Farmers' Gazette* reported on a competition in Fingal, County Dublin. I don't know how raucous ploughing matches were back then, but the reporter noted that 'the assembled peasantry and tenantry behaved themselves with due decorum'. The top prize at that event was an agricultural implement, worth £3. And on 11 March 1864, the *Tralee Chronicle* reported that the Abbeydorney Ploughing Society had its eighth annual meeting in the previous week.

National Ploughing Association trustee Tom Fahey, from Galway, has told me about a contest in Kilkenny in 1868 when the Kilkenny Hunt offered a silver challenge cup for the best ploughmen. A keen historian, Tom believes this is the oldest cup awarded for ploughing in Ireland.

So, between 1816 and 1930 ploughing competitions thrived between neighbouring parishes and the support was enthusiastic, but no one had thought of holding one national contest. Anyway, JJ and Denis decided to have a competition to see which county would come out on top. They didn't know it at the time, but they were making worldwide history as Ireland was the first country in the world to hold a national ploughing competition.

Both men were Fianna Fáil politicians and, in the back of their heads, perhaps they were also thinking that this would be a good event for their rural communities. In 1931 it was a bleak time in rural Ireland, and indeed all over the country. Living conditions were still very poor, most farmers lived on very small holdings and there was a worldwide depression. The Republic of Ireland was only getting over the impact of the War of Independence and the memories of the Civil War were still fresh in people's minds. The Minister for Agriculture, Paddy Hogan, who held that office from 1922 to 1932, had urged farmers to get 'one more sow, one more cow, one more acre under the plough' in a bid to increase productivity, but progress was slow. So, God knows, the country needed a lift and perhaps a national ploughing competition was the thing to do just that.

JJ and Denis were well-known, so they had no difficulty generating interest in this competition. Word spread and soon nine counties had taken up the challenge. The organizers were thinking ahead and they had already decided this was not going to be a once-off event. They declared that the winning county would host the contest the following year. They also secured a Perpetual Challenge cup from the Estate

Management and Supply Association, to be presented to the winning ploughman. A second trophy, the David Frame Cup, was procured for the winning team. (David Frame owned the Hammond Lane foundry in Dublin and he had a strong interest in ploughing.) Both cups are still being presented every year and have stood the test of time and still look magnificent – a testament to how well made they were. We always bought our trophies in Hopkins & Hopkins jewellers, which was beside O'Connell Bridge in Dublin city. That shop was commissioned to produce the original Sam Maguire cup in 1928, so we were in good company.

The competition got underway amid rain and sleet on Monday, 16 February 1931 in a field in Coursetown, Athy, provided by Captain WK Hosie. It was nice and fitting that we were able to return to that exact site in 2011 for the eight-ieth anniversary of the start of the Ploughing. That land is now owned by Jim and Iris Fox.

It cost £9 3s. 5d. to run that first event, and we still have the little catalogue they brought out. It cost 3d. and had advertisements for businesses such as Hogg and Roberts-town seed providers on Dublin's Mary Street, Garryowen Plug tobacco and Ransome's ploughs on Great Strand Street, Dublin. The advertisement for the since-departed Central Hotel on Leinster Street in Athy boasted that it had baths with hot and cold water, and electric light throughout. Down the street, Jackson Brothers, which was a one-stop shop for everything, was advertising its 'high class groceries' along-side seeds, manures and galvanized iron.

Carlow County Library has the old copies of the *Carlow Nationalist* on microfilm and they located the article about that first national event for me, from the 21 February 1931 edition. I had forgotten, until I read it, that local ploughing matches had been held in Athy for the previous three years. JJ and Denis must have thought this would make Athy an ideal fixture for their national challenge. The *Carlow Nationalist* reporter dubbed the event 'The Battle of the Ploughs' and commented that the inter-county challenge had greatly enhanced the importance of the fixture.

The three best ploughmen in each of the nine counties took part in the contest. The counties that entered were: Carlow, Kilkenny, Offaly, Leix (as Laois was then known), Kildare, Wexford, Wicklow, Dublin and Cork. While Louth signalled its interest in entering a team, no team members are listed in the catalogue.

I came across one written account of the event in our records that was a bit disparaging about the quality of the ploughing, saying it was 'variable to say the least, and left room for improvement'. The *Carlow Nationalist* was more generous, saying the judges were unanimous about the excellence of the work done and that one judge had said his task would be 'severe' because of the keen competition. Some 3,000 people attended the event and by all accounts they refused to let the bad weather dampen their ardour.

Denis Allen won his wager on that occasion – the Wexford ploughmen were the best in 1931. The Wexford trio of Martin Conroy, Michael Redmond and Edward Jones won the David Frame Perpetual Challenge Cup as well as a cash prize of £12, or £4 per ploughman. It was a good day out for Edward Jones, who also became the first individual winner of the National Ploughing Championships. He got a cash prize of £5 for his troubles. To this day, Wexford still has extraordinarily good ploughmen. It's like a religion for them down there, like hurling is for Kilkenny people.

JJ and Denis also had some special prizes on offer, the best of which was surely the 10 stone bag of flour, presented by Mr J. Gracie of Kilmeade, Athy. That prize was for 'the married competitor with the greatest number in family'. We don't have a record of who won that prize, but no doubt the hefty bag of flour fed many mouths in his house. That would have been a great prize at the time. Other prizes included a gold medal, valued at £5, for the competitor doing the best work in an inter-county class, and a cash prize of £1 for the best work done by an Irish-made plough.

I often think it would be a lovely idea to gather all the ploughing memorabilia and open a little museum some-where. We would have no bother filling it, if we put out the call to our ploughing folk. I was thrilled in 2016 when President Michael D. Higgins's wife, Sabina, asked me to find a horse plough for a special project in Áras an Uachtaráin. She was designing a garden with the theme of *The Plough and the Stars*, in a nod to her theatrical background. Louth man Gerry King, who is a renowned national horse ploughing champion, presented her with a 1930s Wexford Star plough in early 2017 and, as I'm writing, the sculptor John Behan – a distant relative of my husband – is preparing the plough. I'm sure the garden will be a fine legacy of the tenure in the Áras of Michael D. and Sabina Higgins.

Anyway, the National Ploughing Championships had been inaugurated and the seed had been planted for what would grow into the biggest outdoor event in Europe. The National Ploughing Association was set up, committees were formed in the competing counties and it was onwards and upwards from then on.

As the years continued, more counties got involved and now every county has plough men, and a few have plough women, in the contest. JJ Bergin forged ahead with gusto, while Denis Allen gradually took a back seat in the organization. Denis's son is Lorcan Allen, the former Fianna Fáil TD who briefly served as Minister of State at the Department of Agriculture in Charlie Haughey's 1982 Government. Denis died in 1961, three years after JJ passed away.

JJ was a great man for keeping records and we have the NPA's minutes books going back decades. I have often spent hours looking through them, admiring his lovely handwriting. There's an entry in the minutes from March 1950, noting that the NPA council agreed to pay £1 per meeting to members who had attended more than two meetings per year if they had travelled more than ten miles to the meetings. They got ten shillings if they travelled less than ten miles. No gravy train there!

Once JJ had put the Ploughing on a firm footing here, he worked with a Canadian and an Englishman to set up the World Ploughing Organisation in 1952. Canada hosted the first World Ploughing Contest in Ontario the following year, and eleven countries took part. It was won by a Canadian competitor, James Eccles. Our records show that JJ and two competitors – Ronald Sheane from Wicklow and Thomas McDonnell from Louth – attended that event in Cobourg. They got £100 pocket money between the three of them before they set sail on the *Empress of Australia*. The trip lasted one month and one day. Now they didn't have to pay fares or hotel bills but still, they weren't going to lose the run of themselves with £33 each for a month abroad. I wouldn't think they had any money left over to return to the NPA.

The first National Ploughing Championships I attended was that held in Cahir in February 1954. I had never been to a ploughing match before and I really didn't know what to expect, but I was quite excited at the prospect. I was also very nervous because I was a shy country girl. I wasn't like a city girl who would be well used to being out and about, meeting everyone. Meeting so many people was a bit of a nightmare for me, and I was a total novelty for them because it was rare to see a young woman at the Ploughing back then.

There weren't many attractions to draw a crowd to the Championships in 1954. It was all about the ploughing. There would have been a marquee with tea and sandwiches, but that was the height of it. Not even a slice of cake. I had to compile the results that evening and it was a daunting task because I had scrutineers at my shoulder, checking to make sure I didn't make a mistake. My hands were shaking as I did the job.

We hosted the World Ploughing Contest in Fossa, Killarney, in October of that same year. It was a big coup for a small country like Ireland to host the second World Ploughing Contest. Dances were organized every night in Killarney and it brought great business to the town after a quiet enough summer. A small cannon was supposed to fire to start and finish the event, but it failed to fire on both occasions. Given that the inscription on the prize-winner's trophy was *Pax Arva Colat* (Let peace cultivate the fields), the landowner, Lady Beatrice Grosvenor, said it was fitting that the cannon didn't go off and disturb the peace in the fields.

I remember Lady Grosvenor as an absolutely lovely person to deal with. We were all struck by her elegance and her beautiful clothes. She was the granddaughter of the first Duke of Westminster. I only heard recently that she was assistant superintendent-in-chief of St John Ambulance during the Second World War and was highly praised for her work. She was also a member of the advisory committee to the United Nations' High Commissioner for Refugees, so she was a very able woman.

My first big job at the World Ploughing in 1954 was being put in charge of the collection of the money. I remember Austin Mescal and Maurice FitzGerald, who were both young agricultural advisers at the time, doing the gate as well. Austin went on to become chief livestock inspector with the Department of Agriculture, and was president of the Royal Dublin Society. This is the calibre of the people who volunteer for the Ploughing! The records show that we collected £2,380 17s. 7d. on the gates. I remember we were 6d. out when we counted up the money. We were very exact, down to the penny. We collected £961 7s. from the exhibitors.

I recall getting a brown tweed frock made for the occasion and I thought I was lovely. I used to get a lot of clothes made by May Burley, a dressmaker in Rathstewart, Athy. She was great for getting the magazines from America and they would give her wonderful ideas for dance frocks. Things you'd never see anyone else wearing. Anyone who knows me knows that I do love my clothes, and I go to great trouble to have nice outfits for the Ploughing. Now I don't drink and I don't smoke, so I think that's an innocent enough indulgence. But it's not just a frivolous thing. I firmly believe you have to portray a good image when you are representing an organization. It applies equally to the menfolk, and I would have a word with them if they weren't well dressed for a presentation. They would always take the advice well and you'd rarely have to say it a second time because they know they are representing the Association. I remember my sister, Eileen, pulling a man aside one year and getting him to pull his trousers out over his wellingtons to cover them up.

'Now the next time you come, you'll bring shoes,' she told him. 'Anna May will kill you if she sees you on the bandstand with the wellingtons.'

You never get a second chance to make a first impression and your clothes are the first thing that people notice. In the past, I have told dignitaries to remove their wellingtons when they come on stage at the Ploughing. And I make no apology for that.

I think there should definitely be a code of dress in Dáil Éireann. It shows respect for the people you represent. There's no reason to look like the wreck of the *Hesperus*. It's an insult to the people you are representing because you are saying that you don't think enough of them to make an effort.

Unfortunately, not all my clothing choices worked out perfectly. When I was still finding my feet at the Ploughing, I decided to wear a lovely pair of high-heeled grey shoes to the Championships in Nenagh in 1956. I had bought a wine-coloured coat and the grey shoes looked well with it. The heels and toes were leather and they were made of suede in between. Knowing what I know now about the Ploughing, what was I thinking of?

I remember JJ Bergin looking at the shoes and saying: 'Where do you think you are going with them?' It was so frosty that they had to delay the start of the competition by an hour to let the ground thaw out. Anyway, somewhere along the way I got a heel stuck in the ground and there it stayed. I was like Hopalong Cassidy leaving the field. It was such a cold day that one of the ploughing judges told me that he was sitting at a table for nearly half an hour that evening before he could get back the use of his jawbone again, to eat his beef dinner.

Back in those days the Ploughing was held in February,

and occasionally January, but the weather wasn't working with us. After one bad experience too many, in 1960 we brought it on to November. It gradually moved back to when we now have it – the second last week in September.

The year after I left my heel in a Nenagh field, I decided to follow the new fashion and wear my first-ever pair of trousers to the Ploughing. I remember going to Pim's, a large drapery shop on South Great George's Street in Dublin, and buying a lovely royal blue duffle coat and matching trousers. The Ploughing was held in Boyle, County Roscommon, that year and it was a total washout, so the trousers were a good choice. And my shoes were sturdier than the high heels I wore to Nenagh. I was learning the art of being fashionable but practical at the Ploughing.

I have to admit that I've had directors of the NPA collecting outfits for me on the way to the Ploughing. Things get so busy in the run-up to the Ploughing that I don't have much time for clothes shopping. I might have seen a nice outfit somewhere but they wouldn't have it in my size, so I'd get them to hold it in another branch around the country and get one of our directors to collect it and bring it to the Ploughing for me. He might think he's carrying a very important package, like a national flag or something, and sure it's only a dress for Anna May.

4. The end of an era

JJ Bergin had some very firm ideas and he was not for turning on them. He came to international attention in 1954 when he had a major falling-out with the World Ploughing Organisation (WPO) over some of those views. This was just two years after he had helped set up the organization. Some twenty-six competitors from thirteen countries took part in the second World Ploughing Contest in 1954, so interest in the world event was growing. JJ Bergin was vice-president of the organization at this stage.

I should explain that JJ was a strong nationalist. Before he became a Fianna Fáil councillor, and before he ran for the Dáil for the Independent Farmers and the Farmers' Union, he had a dalliance with Sinn Féin. In July 1917, the *Carlow Nationalist* reported that JJ Bergin had announced that Sinn Féin was the only hope for Irish nationalism because the Government had 'humbugged' the people for too long. But he said that while he believed in constitutional Sinn Féin, he could not have anything to do with a policy that stood for open rebellion.

That nationalistic stance might explain why he took a strong position when the WPO allowed Northern Ireland to send two competitors to the first World Ploughing Contest, in Canada, in 1953. The rule was that every country was allowed to send two competitors so JJ objected, arguing that Ireland was not two countries. He said the WPO decision suggested that it approved of the partition of Ireland. JJ did not get backing from other board members to fight his case, but before that first World Ploughing Contest got underway, he tried again. This time the National Ploughing Association offered to withdraw a competitor to make room for a Northern Ireland competitor and the two chosen would both represent Ireland. The offer was rejected and Northern Ireland sent a competitor to Canada.

JJ was very unhappy about this outcome and the dispute rumbled on into October 1954, when Ireland was hosting the World Ploughing Contest in Killarney. Two Northern Ireland competitors were competing and, to add fuel to the fire, one of them, Hugh Barr, won that world contest. This brought the matter to the fore and JJ felt he couldn't ignore it any longer. At a meeting of the NPA's council on 9 November 1954, he secured unanimous agreement to remove the National Ploughing Association from the World Ploughing Organisation. He was clearly very annoyed about the whole thing because he was quoted on the front of *The Irish Times* saying the 'small clique' at the head of the WPO had rejected several gestures he had made to resolve the matter. On another occasion he declared that his Association 'would not prostitute Irish national ideals'.

I remember being very disappointed that there was such an abrupt finish to our involvement in the World Ploughing Organisation. It had been wonderful to be a part of a world event, albeit for a very short time. While the visiting competitors were in Ireland they were brought on various scenic tours. Before leaving, they went to the Irish National Stud and Japanese Gardens, afterwards lunching in the famous hotel of its time, Lawlors Hotel in Naas, County Kildare. JJ got a lot of criticism for removing the NPA from the world body because our involvement in the World Ploughing

Contest was an exciting opportunity for our best plough-men. But he was also commended by others. I remember many letters coming in, praising him for his stance. Other letters asked if we'd ever return to the WPO.

After the furore, the WPO changed the rules to say that every affiliated ploughing society could enter two competi-tors. As Northern Ireland had its own society, and we had ours, that removed the politics from it. We didn't like to think that we were depriving our competitors of the chance to win a world title, so after discussing it at our national council we rejoined the WPO in 1960. Republic of Ireland and Northern Ireland competitors have been competing alongside each other ever since, and our two organizations get along very well indeed. Sadly, JJ was no longer with us when we rejoined the world body, but I'll tell you more about that later.

JJ was always thinking of ways to improve our own plough-ing event and in 1954 he took the notion to introduce a ploughing class for women. Now, to be honest, I don't think JJ was proposing it for feminist reasons. He was thinking of it as a publicity stunt, and it definitely worked in that regard. It cre-ated a terrible amount of excitement at the time. Some people thought that the wise old owl was deteriorating with age when he came up with this plan. He was seventy-four at the time.

Reading the documentation from that time, it sounds like they thought that he had gone mad, to be honest. You have to remember that this was conservative Ireland in the 1950s and the NPA's council was a very male, traditional group of people. They were set in their ways. In the early years, long before I got involved in the Ploughing, the annual dinner was a men-only gathering. I suppose it wasn't too hard to have a men-only dinner when no women were involved!

The men who opposed JJ Bergin's plan seemed to

overlook the fact that many women farmed and ploughed. In the Ireland of the 1950s, both men and women did a lot more physical work than they do on farms now because they had none of the modern farm machinery that we take for granted nowadays. But, as well as that, women had none of the modern conveniences inside the house either. These farm women wouldn't be daunted by a plough.

But back in the Ireland of 1954, it was a different story for women who wanted to plough competitively. There were ructions over JJ's plan to introduce a female ploughing class. Former journalist and public relations guru Larry Sheedy recalled it well when he compiled *65 Years of Ploughing for Progress*, a history of the NPA, in 1996. Larry was, and still is, a great friend to the Ploughing. His father, JJ, was the Dublin representative of the NPA. Larry wrote that 'women's lib' hadn't reached the council of the NPA and the idea of a class for women was seen as revolutionary. People thought JJ Bergin had 'gone over the top', was 'too gimmicky' and – the best one – 'was introducing Hollywood razzmatazz at its worst'. Now, I defy anyone to watch the women ploughing at any of our events and find a bit of Hollywood glamour. They'd be hard-pressed to find any razzmatazz, but they would find the best plough women our country has to offer. I was watching a novice class of three girls and three lads competing recently and the girls were wonderful. One of them drives a JCB at home and they say she can turn it on a sixpence.

Despite the ferocious opposition to his plan, JJ persevered and he got his way. He decided that the winner of the competition would be known as the 'Queen of the Plough'. Weir & Sons of Grafton Street were known as the leading silversmiths in Dublin at that time, so off we went to Weir's for a suitable award for a champion plough woman. We bought a beautiful silver crown with a Celtic design. It cost £500 then and was recently valued at €7,000, so it's a valuable piece of our history.

The first farmerette class was held at the National Ploughing Championships in Cahir in 1954. Elizabeth Murphy, better known as Cissie, from Ballagan, Cooley, County Louth, won the title. You wouldn't find that name in the catalogue, though, as she simply entered as Miss Louth. She wasn't the only woman representing Louth in the inaugural event – A Miss B. Lawless from Kilcroney, Ardee, also entered in the class. Cissie Murphy was very active with Cooley Macra na Feirme, and I believe there were great celebrations in County Louth when she took home the crown.

The crown went as far south as you could go from Louth the following year when Anna Mai Donegan from Kerry took the title. She retained it in 1956.

JJ Bergin was an inspired man when it came to generating publicity. He must have known that the competition would really catch the public imagination. The winner was announced at the Queen of the Plough dance at the end of the Ploughing, and that event was always absolutely thronged. Before the dance, the winner would be whisked off to the nearest drapery shop, where she would be provided with a lovely frock. The drapery shop always donated the frock and they would have put it on prominent display before the Ploughing, to generate a bit of interest. A nip and a tuck, and the Queen of the Plough was ready for the ball. She would be kept out of sight until the crowning. This event was the climax of the dance and it always took place at midnight. We usually had some dignitary on hand to do that and there would be huge excitement in the ballroom. The crowds would be hanging out of the rafters for that event.

In his wisdom, JJ Bergin deemed that the competition would be called the 'farmerette' class. Before you start blaming him, he didn't invent the word. The *Collins English Dictionary* has an entry for farmerette, describing it as a woman or girl who works on a farm. It also offers farmeress, which JJ could have chosen. But he went for farmerette, and it's a word that is always very controversial with the media.

Journalists often ask me why we don't just get rid of the name, saying the 'ette' belittles the women taking part. Funny enough, the women themselves never question the title at all. I don't think it's demeaning. I've always said it's not for me to change it and I don't think it will change. It's part of the history of the Ploughing. I wouldn't like to take away something that the founder of the Ploughing started. It's a part of our history now. I just want to enhance what's there.

Would you believe that I even had to defend the title of the competition in a debate on national television many years ago with representatives of the feminist movement? I stuck to my strong belief that JJ Bergin actually promoted the cause of feminists by introducing the competition at a time when the term 'women's lib' was never even heard of.

It's just as well the dowry scheme is long gone or the feminists would go mad! That was another one of JJ Bergin's brainwaves. He came up with a plan to give a dowry of £100 to the Queen of the Plough if she got married on or before her twenty-fifth birthday. It was claimed only once in our history. Mary Shanahan, now Mary Falvey, from Abbeydorney, County Kerry, was the lucky woman. She won the title in 1959 and 1960. The year of her first win, 1959, was a unique year for us because we held two National Ploughing Championships. We were finding it increasingly difficult to cope with the February weather during the Ploughing, so it was decided that we should move the event to the late autumn. To facilitate the changeover, we had the Ploughing in January of that year, in Burnchurch, County Kilkenny, and then in Oak Park, County Carlow, in November. Mary won the Kilkenny contest, and Muriel Sutton from Wicklow won in Oak Park. Mary still comes to the Ploughing and she hasn't aged a day. All that ploughing must be good for the skin. She made a second bit of history with her father because they were the first father-and-daughter team to win a national ploughing title. P. J. Shanahan won the championship in 1951.

I'll tell you something now that demonstrates what a small country Ireland is. In 1957, two women jointly won the Queen of the Plough title – Muriel Sutton again, and Eileen Duffy from Galway. Eileen married an army man and it was their son, Mr Hugh O'Connor, who delivered my daughter Anna Marie's baby, Saran, in the Coombe hospital in 2010. I'm constantly amazed at the way people you meet through the Ploughing pop up in your life in ways you'd never expect.

It was a very proud day for my parents, and myself, when my youngest sister, Eileen, won the Queen of the Plough title in 1961. We were thrilled for her. The Ploughing was in Killarney that year and my parents travelled all the way down for the Queen of the Plough dance because it was such a great achievement. Then she went ahead and won it two more times, in 1963 and 1964. But that wasn't the only victory she had at the Ploughing. Eileen won the National Brown Bread-Baking Competition in 1973. That was the last year of the bread-baking competition before it took a long break. Then it was revived forty-one years later, in 2014. Eileen was delighted to be asked back as a judge in that competition. She has been judging it since then, and will be back judging it again this year.

But back to the farmerettes, because I should also point out that my daughter, Anna Marie, managed to take the Queen of the Plough title in 2008. We always said Anna Marie would have to win it very well to get the top prize because people would say she got favouritism because of me.

It's hard to believe now, but the NPA had the honour of leading the St Patrick's Day Parade in Dublin in the late 1950s and early 1960s. That shows how important agriculture was in the country at that time, doesn't it? You couldn't imagine it happening now. When the Queen of the Plough was initiated, it was agreed that the winning woman would lead the parade on a new tractor. She was feted everywhere. The winners of the individual competitions followed her, also on new tractors provided by the manufacturers. Then the format of the parade changed sometime in the 1960s and we were moved to the back of the parade, so we pulled out at that stage.

Several women have entered the World Ploughing Contest alongside men over the years, and have performed with great distinction. Helga Wielander from Austria was the first woman to become a World Ploughing champion in 1993. She won the conventional competition, beating fellow Austrian Josef Gadermayer into second place and Ireland's Eamonn Tracey into third. It was a big deal in Austria and one ten-year-old girl was captivated by the story. Little Barbara Klaus dreamed that one day she might also become a world champion. Barbara Klaus recounted that story in 2013 when she became a World Ploughing conventional champion. She said Helga Wielander was her biggest idol and she was thrilled to win the same competition exactly twenty years later. It was a good year for Austrian women as the second Austrian representative, Margareta Heigl, was runner-up to Wexford man John Whelan in the reversible ploughing competition. Now, I can tell you, when the women took gold in 1993 and 2013, some of the men were not a bit happy. They couldn't believe that the ladies could beat them. But the women were treated exactly like the men. They got no extra assistance and they definitely deserved their victories.

I was most certainly a woman in the men's world when I started working with JJ Bergin. He set up the NPA as a limited company in 1956 and made me company secretary, while he was managing director. As well as my normal duties, my new role involved keeping company records, giving details of our directors to the Companies' Registration Office and updating that information every year. I have to say, he had great foresight when he appointed me because women were not to the forefront at all in the business world at that time. I have no doubt JJ would have got a lot of criticism for appointing me to such a crucial position. People were openly saying: 'Look, can a woman do this job?' Women were always in the background and if there was a top job going, men always got it. But once JJ made a decision, he stuck with it, and he had decided I was fit for the job. And the NPA members were fully behind it. My appointment was noted in the press, and I'm sure some people were surprised to see a woman getting a job like that.

My maiden name was Brennan and when I answered the phone, people would automatically ask to speak to Mr Brennan, thinking I was just there to answer the phone. I would have to really persuade them that there was no Mr Brennan. I remember one guy saying to me: 'Now what on earth would you know about ploughing?' I simply said to him: 'Just ask me the question, please.' And of course I was able to answer it. But that was a reality for me and it was quite challenging because I was young at the time and didn't have the confidence I would have now. I don't remember reflecting on gender inequality at that time or having the sense that I was in the vanguard of women who were rising in the ranks. I was just focused on doing my work to the best of my ability. And I knew I had the support of the NPA council. They didn't care if I was a man or a woman as long as I did the job right. They were good times, though. In 1958, I was earning £5 a week. You could do a lot with £5 in those days.

Back then, I remember packing all the paperwork we needed for the Ploughing into a small school case. That wouldn't hold our pens and notebooks now. We have horse boxes and trailers, containers and crates of paperwork carrying all our equipment to the Ploughing.

As I got more familiar with the work, JJ Bergin's health gradually began to fail. It was sad to see someone who was so strong and vibrant getting weaker. I wonder if he knew that he was on the way out? I say that because we had the Ploughing in Tramore in February 1958, and he said something that stuck with me when he stood up to speak at the Queen of the Plough dance. He said he hoped he would be leaving behind a working machine when he went from the Ploughing. A few weeks later, he had a stroke.

I'll never forget the day it happened. We were going into Athy to post letters and he was waiting in the car for me with the engine running. I sat into the car and said I was ready to go, but there was no answer. When I looked over at him, I knew something was wrong. I got out and ran around the car and opened his door. His arm fell down and I ran for help.

He never regained consciousness after the stroke and he died at home eight days later, on 13 March 1958. He was seventy-eight years old. And hadn't he packed a lot into those seventy-eight years and left such a legacy? His passing was like a death in the family for me. He was really like a father figure for me and was a great adviser. More than fifty years later, I still remember some of the advice he gave me and it has stood the test of time. For example, he always said never to sign anything without reading it first, and I've always stuck fast with that rule.

The NPA office was in his house and it was very hard to return after the funeral and face his empty chair. I remember going back to work the week after he passed away and it was very lonely, terribly quiet. Eileen came with me to keep me company for the first few days because I don't think I could have faced it otherwise. I knew the whole family so well at that stage and we were all grieving together. But getting back to work helped, because it gave me something to focus on. And of course I was still hearing JJ's voice in my head. I would often ask myself, 'What would JJ do?' when faced with a decision.

After JJ Bergin died, National Ploughing Council chairman Michael T. Connolly phoned me to ask me to run the show until they appointed a new managing director at the AGM. After that meeting, JJ was replaced by Seán O'Farrell, an All-Ireland medal-winning hurling star from Ballyhale, County Kilkenny. I must say I was very pleased with the appointment because we had a lot in common, not least sport. Between Seán's hurling and my camogie playing, we never ran out of conversation. Seán had won his All-Ireland medal with Kilkenny in the 1933 final against Limerick, although he was a non-playing substitute and was carrying an ankle injury from the semi-final against Galway. That Kilkenny team also had the distinction of being the first Kilkenny team to have won the All-Ireland championship, the provincial championship and the hurling league in the same year.

Seán had later moved to England for a time and he'd played cricket for Surrey, so we also had that sport in common. When he returned to Ireland, he met Lil Doyle, of the famous Lil Doyle pub in Barndarrig, County Wicklow. They married and Seán lived there for the rest of his life. Not surprisingly, he got very involved in the Wicklow GAA, serving as chairman of the county hurling board. He also campaigned to lift the ban on GAA players playing non-Gaelic sports. I'm glad he lived to see the ban being overturned in 1971.

Seán was very easy to work with and so helpful, always at the other end of the phone if I needed him. At this stage the office had moved from JJ Bergin's home to the loft in my home place of Clonpierce, but Seán did his work from Wicklow. Seán O'Farrell was very different from JJ Bergin. JJ was larger than life, while Seán was a quiet, calm character. He wasn't small, mind you. Seán was over 6ft tall, but one man towered over him. That was the French president, Charles de Gaulle. Seán was very proud of a photo he got with de Gaulle because Seán looked like a small man beside the 6ft 5in tall French president. They met when we went to the World Ploughing Championships in France in 1961.

He was at the helm for another Ploughing first – RTÉ's arrival at the event. It was in November 1961 that the broadcaster attended its first National Ploughing Championships, which were held in Killarney. I understand it was one of their first shows, if not their first, to be filmed by their outside broadcasting unit. The footage was used for the first episode of a farming programme called *On the Land*, which was broadcast on 1 January 1962. Why the two-month delay? Well, Teilifís Éireann only went on air the day before the broadcast. In it, presenter Noel O'Reilly talked about the poor weather at the Ploughing – nothing new there – and said the land was a little bit bumpy for top-class ploughing. As it happened, I went to school with Noel's wife, Ida Reddy. She lived two miles from me, a lovely, lovely girl. Noel also interviewed Lady Beatrice Grosvenor, chairwoman of the local organizing committee, who said she had been busy organizing horses and accommodation for the competitors. She said she found it much wiser to sit back and let everyone else do the work as efficiently as possible. Sensible lady.

5. John

Ballylinan Hall would not be known as the place where great love stories begin, but happily for me it played a part in bringing me face to face with a young man called John McHugh. I first remember seeing him at a Muintir na Tíre meeting in the hall in 1954, when I was twenty years old. The community council was very popular in Ballylinan and John was chairman for a few years.

What attracted me to him? I don't know. There was just something about him. It's a bit of a cliché, but he actually was tall, dark and handsome. He was a good chairman, firm but not too serious. There was something jolly about him. He was always a calming influence if there was anything contentious going on. Like me, John was very involved in the local community. He was in the Catholic Young Men's Society and the Pioneer Total Abstinence Association. The simple things of life then, that young people wouldn't even know about now. These organizations were very important in country areas. They were a social outlet and they meant that everyone knew everyone.

I got to know John well from our Muintir na Tíre meetings and I thought very highly of him, so I was delighted when he asked me out. He confided afterwards that he'd had his eye on me from afar for some time, which was news to me. Our first date was at the Ballylinan carnival dance. The carnival happened every few years and it was a great attraction for people of all ages. The amusements were very exciting – everyone loved the chairoplanes – and I remember how we used to wait around until midnight to watch these fellas climbing a pole. They were dressed in white and they would shimmy up the pole and do acrobatics at the top of it. It was a big attraction. The carnival dance was always packed and it was really the place to go for the music and the dancing.

John was from a farm on the other side of the village, about three miles from our house. He was one of six children. They were a very devout family – two of his sisters became nuns, something that would be unheard of nowadays. His sister Kitty became Sr Dominic with the Sisters of Mercy in Athy. She was matron with St Vincent's Hospital in Athy for a long number of years. His sister Eileen joined the Sisters of St Joseph of the Sacred Heart order. She became known as Sr Kilian and went to Australia with the order, to train as a teacher. It must have been very lonely for her because she had just left school at the time. When she set sail from Dún Laoghaire, she stood watching her mother on the harbour until she was the size of a pin. Eileen never saw her mother or father again because she was in Melbourne for thirty-three years without ever again getting home. That was very tough on her, cruel in fact. Both parents died while she was out there. After a nun was notified that a family member had died, the sisters would pin a notice on the board asking people to pray for the repose of the relative. I can't see the point in denying people the opportunity to see their parents before they pass away. What difference would it have made to allow them to go home to see their families? Both Eileen and Kitty have passed away since, and John's only brother, Richard, and his sister Mary Ann also died some years ago. Brigid, John's only surviving sibling, is still living in Athy.

I have lovely memories of the time when myself and John started going out together. We'd go to the pictures in Carlow and occasionally a group of us would go up to the Ambassador cinema in Dublin. The dances in the Ritz ballroom in Carlow on a Wednesday night were always a good night out. Then once a year there was the Laois Farmers' Dance and the Kildare Farmers' Dance. They were magnificent nights out, with great style. John was a lovely dancer, a great waltzer.

Myself and John went out for a good number of years before we settled down and got married. I can't gloss over the fact that I was thirty-two when I got married, so it was a long courtship. Gay Byrne gave me an awful time about this when he interviewed me for the *Meaning of Life* programme on RTÉ in 2016. I remember him saying: 'You kept him waiting for *how* long?' And in the end, I said something to the effect that I was obviously well worth waiting for.

The truth was that John was living at home and his brother, Richard, was going to take over the home farm, so we had to find somewhere to live. Back then, you didn't get married until you had a home to move into. I remember a lady saying to me: 'John has the bird, but he doesn't have the cage.' I thought that was very apt. I was also busy with work and, to be honest, I wasn't in a big rush to get married.

John's uncle owned a farm locally in Fallaghmore, near Ballylinan. He had no children and when he died, his wife sold the farm to John. The way was clear to get married then, and John proposed to me outside the hall in Ballylinan. He didn't get down on one knee, though. He had already gone to ask permission from my father, so it was no surprise when he popped the question. I remember when he arrived at the house, asking to speak to my father. Poor mother was left behind. She wasn't asked at all. It was always the father who was asked. My mother loved John, though, and they were always very close. It was just the way things were done at that time. When he arrived, John was brought into the parlour and a few of my brothers were listening at the door. John was like a brother to them at that stage. I asked him afterwards what Dad had said to him and it was something along the lines of: 'It took you long enough. I thought you'd never ask.'

Then we went along to Hopkins & Hopkins jewellery shop on O'Connell Street in Dublin to get the ring. You might remember that shop from earlier, when I mentioned how we got the first cups for the Ploughing there. The NPA had a long relationship with the shop so when I thought about picking an engagement ring, it was the obvious choice. Back then the wedding was planned with your parents and held to suit them because they were paying for the wedding breakfast. We got engaged in September 1965 and were married in the Sacred Heart Church in Arles on 28 July 1966. That time was picked because it suited the farm work at home. If we had to plan a wedding now, we'd be taking ploughing dates into account, rather than the farm work.

We had our wedding breakfast in the Downshire Arms in Blessington, County Wicklow. Then we went off to the Aran Islands on our honeymoon. I'll always remember getting a bit teary when I left my mother and father for the honeymoon because I was such a home bird. John noticed and said to me: 'Aren't you lucky to have your parents still with you? I wish I had mine.' He was right. His father died in 1951 and his mother died in 1960, so he had no one to wave him off on his honeymoon.

It was just lovely, travelling around the Aran Islands on a pony and trap. We spent some time in Clifden, too. We went back a number of times since, on holidays with the children.

When we returned from the honeymoon we moved into Fallaghmore and this is where I've lived since, and where I've run the NPA office from for the past forty-one years.

Myself and John were a good match. Now, our personalities were quite different, but they complemented each other. John was calm and steady, very thoughtful, whereas I was more hot-headed and liable to take action on the spur of the moment. We suited each other well.

Like me, he was a lifelong pioneer and, like me, he didn't have any great hatred of alcohol, he just preferred life without it. Don't get me wrong. I'm not a goody-goody. I like a bit of fun and entertainment, but I don't need to have alcohol to do that. It wasn't unusual to be a pioneer when we were growing up. I remember being at a ploughing function and there were five or six of us in a group and not one of us took a drink.

John smoked a pipe, something you don't see much of nowadays. He loved his pipe and you always knew he was content when you saw the pipe coming out. My son DJ used to say he knew he was doing a good job when he was ploughing if he looked over at John and saw the plume of smoke rising from his pipe.

John and I worked terribly hard on the farm when we got married. We had to, because we had to repay the loan we got to buy it. There was an old two-storey farmhouse on the land and we worked hard modernizing that, because it hadn't been done up in a long time. John was a farmer to the backbone. It was mostly tillage and livestock farming he was involved in. He grew sugar beet too, back when we still had a sugar industry in Ireland. And while he never competed in ploughing matches, he was a very tasty ploughman.

I loved the outdoor work. I preferred working with the

sheep because I found them easier to manage than cattle. I always loved getting away from the paperwork and walking down the field, talking to myself if I felt like it. It proved very useful when I was in the Ballylinan drama group because I'd practise the lines as I'd go off down the fields.

There's something special about fields. There's a peace about spending so much time close to nature. I have a love for the land that nothing could replace. It's a happy way of life mostly. But then you see in the *Farmers' Journal* that subsidies are being reduced or there's some new rule from Brussels and a bit of the peace ebbs away.

In the same way, I feel a great serenity watching ploughing. A lovely peace descends on you when you're watching the sod being turned to face the sun and the seagulls following the ploughman as he methodically goes up and down the field. I think you're very near to God when you're on the land.

Patrick Kavanagh puts it very well in his poem 'Ploughman', published in 1936, when our Championships were only getting going. He talks about painting the meadow brown with the plough and how 'Tranquillity walks with me/And no care./O, the quiet ecstasy/Like a prayer'. Lil Doyle, the late wife of Seán O'Farrell, our second managing director, had a copy of that poem hanging on her wall and gave it to me many years ago after I admired it. I think it really highlights the connection between the land and God, and in a more poetic way than I could ever do.

There's a great peace working in the garden, too. I'd often stay out in the summertime until half-ten at night. When you come in, you feel like you've truly achieved something.

DJ was born in 1967, the summer after we got married. I remember I had to be at an executive meeting in Tullow on the Monday night, but DJ had other plans and I ended up in

St Brigid's Hospital in Carlow instead. DJ was born the next morning. There was such joy over his birth because it was the first grandchild in the Brennan family. My father insisted on buying the pram. But there was no way it would be brought into the house until the baby was born. I remember he went into Athy and got this lovely, high, navy-blue and white pram. Those big prams are back in fashion again, I see. My sister Betty was a great support when DJ was born. She would come to the house every evening after work and help me. Her nursing training shone then, because whatever way she put him down at night, he always slept soundly.

I remember thinking that the house would never be tidy again when DJ was born. There were baby clothes every-where. And then Anna Marie arrived two years less a week after DJ. I didn't know it then, but our little family was com-plete. I would have really loved to have had more children, but it just wasn't to be. I would have loved a little sister for Anna Marie. Still, I always say I was blessed with two lovely, healthy children so I can't complain.

Of course, there was no maternity leave when I had my babies, but then the Ploughing wasn't as big as it is now. Nevertheless, it was a busy time. I was lucky because my sisters Eileen and Betty were such a good support when the chil-dren were young. And, of course, John was always there. He was always a very caring, loving father with a wonderful temperament. He was great at Irish, but the teachers always knew when he helped DJ or Anna Marie with their Irish homework because the word would come back: 'Tell your Dad it's good, but it's the old Irish he's teaching you.'

Our home was what we used to call a rambling house. People would drop in unannounced all the time. That was down to John, nothing to do with me. He had a great way of drawing people to him. But it wasn't only people. If John was out on a summer's evening painting a fence, every bird for miles around would be sitting on the fence beside him and he'd be surrounded by cats and dogs, even the geese. And maybe a peacock, too. We kept peacocks for a while and it's not for nothing they have a reputation for vanity. I remember one peacock used to perch on the sitting-room windowsill for hours on end. What was he doing? Looking at himself in the mirror over the fireplace.

John's favourite perch was in the kitchen, pulling on his pipe and telling stories. He was a great storyteller and people would congregate after a football or hurling match to hear what he had to say. He loved telling stories that he had heard when he was a young lad. He'd put a shiver down your spine with some of the stories about ghosts. There was the night he was cycling home from a dance at Vicarstown, County Laois, and he had to pass the gates of Ballyadams Castle, a fifteenth-century house that's in ruins now. It was a lovely moonlit night and as he and his cousin parted ways, the cousin told him to look out for the ghost at the castle gate. John was getting nervous as he approached it and suddenly he heard a rattling sound. Was it chains? Or the big gates being yanked open? He froze, and then decided to pedal like hell until he passed it by. The rattling grew louder and as he passed the gate he looked out of the corner of his eye. There was the culprit – a big old bullock scratching himself on the gate.

He had plenty of stories about wakes and banshees. Younger readers have probably never heard of the banshee – the wailing woman with the long hair who appears to people just before a family member dies. She is always combing her long hair. As youngsters, we used to be warned against picking up a comb off the ground in case it belonged to the banshee. By picking it up, we were bringing her into the house to reclaim it and thus causing the death of someone in the family.

John used to talk about the appearance of a banshee at a wake up the road. A group of lads was gathered on both sides of the coffin. Then suddenly the boys on one side went white in the face and started staring behind the people opposite them. They all said they could see a banshee standing outside the window, combing her hair. I can't remember hearing if someone died soon after but I presume they did, or the story wouldn't have been retold so often. People really did believe in ghosts and banshees back then and it's understandable because there was no electric light. When you are sitting in the dim light of a lamp, it's easy to let your imagination run away with you and to hear or see things that aren't there at all.

John was a great follower of sport, but sometimes the excitement got too much for him. I'm thinking specifically of the time he was watching the Leinster football final with my sister Eileen's husband, Tommy, in 2003. Laois were up against Kildare and they are always great rivals. Laois hadn't won a Leinster title since 1946, so it was a big deal. Halfway through the match John and Tommy couldn't take the tension anymore so they covered the television with a throw – that made no sense at all – and they went off down the road to look at a new house that was being built. Of course, their minds were on the match and they didn't take in anything when they looked at the house. Anyway, Eileen came after them to say that Laois had won by a goal and they came flying back to the house to see the celebrations on the television. When it was over, John said to Tommy: 'Will we go back now and look at the house?' It was as if they hadn't seen the house at all, they were so overwrought about the match.

That sense of excitement would be exactly the same for people who have a passion for ploughing, but some folk would find that hard to believe when they stand watching the ploughmen moving up and down the plots.

I could talk all day about John, I thought so highly of him. He was a man before his time when it came to the children. I never had to ask if he would collect them from school, prepare a meal or take them to their various social outings. He was just a very hands-on dad. John was a very mellow man around the house and it would take a huge amount for him to raise his voice, but I can assure you that when he did raise his voice, the kids knew he was serious.

John never got fussed or pressurized so we were quite different on that score, but somehow he always seemed to get through the workload. He came from a generation that didn't know their own strength and never took the short cut around any task. Family, friends and neighbours meant everything to John and he never spared the time he gave to others, always stopping his work when someone came to him in the field or up the haggard for a chat. He had an uncanny way of judging people and he could sum a person up very quickly. There were absolutely no airs and graces about John and while he would never shower me with compliments on how I looked or how I was getting on with the job, you always knew he was proud and approved. When the media might print a negative story about the Ploughing he was often heard to say: 'Have they nothing else to be writing about?'

John was such a faithful Irishman that he would not even go abroad on holidays, saying: 'Sure what would be over there that could be any nicer than what we have here?' He just loved the Irish countryside. The nieces and nephews adored him too and I think they would all have memories of summers spent on the farm making ricks of hay, swimming in the big rainwater barrel, sitting on the top of the bales on the way up from the fields or learning to drive the tractor. You wouldn't dream of allowing children to do many of these things now, but times were different back then. John even let them try his pipe. When scolded for that he would say: 'Musha, sure what harm would it do them?'

6. Taking the reins

When I think of the years spent working with Seán O'Farrell in the NPA, I remember a huge element of excitement about the Ploughing. The Ploughing was getting bigger and better every year. We were all growing with it and facing new demands every day. It was a bit of a challenge, and I do love a challenge. I was getting to know people and I felt I was winning their respect. I had found my feet in the organization and, indeed, in several more. I was busy with the Kildare County Show at this point, and then there was Muintir na Tíre and the GAA and all the other local organizations I was involved with. I remember someone saying to me: 'Anna May, you're in everything except the Men's Sodality.' Now that was one religious group I obviously couldn't join! I was, however, in St Anne's Sodality, a women's group. Younger readers have probably never heard of the Sodality. It was a get-together after monthly confession, Mass and Benediction.

Just like JJ Bergin, Seán died very suddenly. After fourteen years working together, I got a phone-call one day to say Seán had taken a turn and had been rushed to hospital. The next morning, he died of a heart attack. He was only sixty-three. That was in October 1972, just before the National Ploughing Championships at Rockwell College, County Tipperary. The sudden death was such a shock at the time because he was far too young. He was as healthy and fit as anyone. Although he was a heavy smoker, like a lot of people at that time.

Seán O'Farrell's stint may have been shorter than JJ Bergin's, but he still left his mark on the organization. He encouraged the policy of taking the Ploughing to the widest possible selection of venues, from Mallow in Cork, to Finglas in Dublin, to Athenry in Galway. He brought in classes for students from agricultural colleges and invited the Irish Countrywomen's Association (ICA) to give demonstrations. He also managed to bring the National Ploughing Championships back to his home county of Kilkenny three times during his reign. And there was no prouder man when Charlie Keegan from Seán's adopted county of Wicklow became the Republic of Ireland's first World Ploughing champion in 1964. We'll come back to Charlie later, and to all those other ploughing legends.

So, for a second time there was a vacancy at the top of the NPA. Just as he had done when JJ Bergin died, National Ploughing Council chairman Michael T. Connolly asked me to keep the show on the road until they appointed a new managing director at the AGM in the following May. He made it clear that I was responsible for keeping everything correct and in order until the opportunity came to appoint a managing director.

Never in my wildest dreams would I have thought of putting myself forward for the job of managing director but, to be honest, I didn't really have a say in it. It turned out that Michael, who was chief agricultural officer in Wexford at the time, had a plan up his sleeve. He called the executive committee to his home in Wexford before the AGM to get agreement on who the next managing director should be. They decided among themselves that I was the person with the best knowledge of the Ploughing, and that Michael would propose me as the next managing director.

I had absolutely no idea that this was going on. It's probably just as well that I didn't because I would only have been worrying about the prospect. Instead, I was preparing for the AGM and wondering who I would get as my new boss. I was hoping it would be someone I could work easily with. It never dawned on me that it could have been yours truly, and I'm being very truthful about that. I was happy in my job and happy with the way my life was going.

So, the fateful day dawned on 10 May 1973 and off I went to the meeting, which was held in 3 Gardiner Place, the offices of the Dublin County Committee of Agriculture. There were representatives from the twenty-six counties at the meeting and I had heard that about six of them were very interested in the job, so I was expecting to hear one of their names being called. I remember mulling over who I would prefer.

Anyway, Michael stood up and spoke about the loss of Seán O'Farrell. Then he said there was a person in the room who was highly qualified for this job, who knew the Ploughing inside out, who knew it better than anyone at the table. I was still wondering which of the six it was and then he said 'she'. I was the only woman in the room, so it didn't take long for the penny to drop. I was shocked because it had never once crossed my mind that I would have been proposed. You could have knocked me down with a feather.

Looking back now, I wonder why I didn't see myself as a contender. I'd say it was because I was a woman and, back then, it was a world run by men. The marriage bar, which forced women working in the public service to resign upon marriage, was only removed two months after I became managing director. Young people might not believe this, but women didn't even sit on juries in 1973. Apparently, there were two justifications for that: their delicate natures might not be able to handle the evidence; and there was a risk that their household duties might be neglected.

In 1973, husbands could still sell the family home without their wife's consent, and women weren't allowed to collect their children's allowance unless their husbands nominated them to do so. And the average hourly pay for women was almost half that of men in the early 1970s. Remember the way people would ask for Mr Brennan, the way they would assume that a woman wouldn't be able to deal with a simple query? I was a product of that time and I suppose that explains why I didn't put myself forward when the opportunity arose.

Men wouldn't think twice about putting themselves forward, even if they hadn't the necessary experience, but as women we were always second-guessing ourselves. I don't do that now. I know what I'm capable of and I hope it's the same for other women coming behind me. I think I've instilled that very well in Anna Marie, anyway.

After Michael T. put my name forward, he was seconded and the rest of the executive agreed with that. No one else put their name forward. I'd say there was terrible disappointment in the room. I could sense it coming from the others who were interested. A strange silence fell on us all as we took in the news. But in fairness to the people who were hopeful of getting the job, they all pledged support to me afterwards and they were as good as their word. I cannot overstate how well I get on with the NPA directors, the county ploughing associations and the competitors around the country. They know they can call me any time and I feel the same about them. I'm a firm believer in the saying: if you want to go fast, go alone, but if you want to go far, go together. I think that's why the Ploughing works so well. I might be the face of the organization and the one to get the recognition and awards, but we'd have no awards without the whole organization. The only awards I can claim for myself are my camogie medals.

When I look back on the day I was appointed, I think that perhaps my appointment caused less friction than if one of the six men had got the job. All the men were eyeing each other up and there was probably a bit of rivalry between them. I was the only woman at the meeting and I hadn't put my name forward, so I think that softened the blow for them. They couldn't say much – well, maybe they did complain about it, but not to me anyway! Now, they weren't interested in the job for the money. The pay was very nominal indeed, but there was a great honour involved.

I remember I was so nervous after getting the job that I could not sign my name at that meeting. I was absolutely trembling. To be honest, I knew as much – if not more – about the NPA as anyone else sitting around the big table, but being managing director was a different level of responsibility. I immediately felt a huge weight on my shoulders. There was always someone else to blame when I was secretary. Everything arrives at the door of the managing director, and now I was that person. I had travelled to Dublin with a man who, I later learned, had been planning to put his name forward for the job. He was very magnanimous about my appointment and said he thought it was a good decision. I can hardly recall the journey home that day, I was in such a daze.

But I do remember coming in the door of my house, both happy and unhappy, and saying to myself: 'I'll never laugh again, I have so much responsibility on my shoulders.' It was very daunting. I was turning thirty-nine the next day and I had two young children – DJ and Anna Marie were almost six and four years old at the time.

John was pleased for me, but anxious, when he heard the news. He asked if it would be a lot more work for me, but in the same breath he said he was there for me if there was anything I needed. I knew he'd be supportive, and he was. He never let me down. The children were too young to realize the significance of it and they just grew up accepting that the Ploughing was a part of their lives.

My promotion to managing director was a very low-key thing and there wasn't much made of it in the media, probably because I had been company secretary for so long and I suppose people knew I could deliver the goods. The shock was greater when JJ Bergin had made me company secretary seventeen years earlier. I certainly didn't get a wink of sleep on the night I was made managing director. I was lying awake, worrying about what lay ahead and wondering if I was right or wrong to accept the job, worrying about all the things that could go wrong. But I was excited, too, at the thought of taking on the challenge. I knew that the Association was in good shape and the future was looking rosy. But I certainly had no idea that I would still be managing director forty-four years later. I really wasn't thinking that far ahead. My sole intention was to give it my very best shot and do the organization proud. I hadn't the luxury of panicking because the World Ploughing Contest and the National Ploughing Championships were coming up in Wellington Bridge, County Wexford, in September, just four months away. I had a four-day event to organize and a mountain of work to do, so it was going to be a massive undertaking. I decided the only thing for it was to put my head down and get on with it.

While it was a daunting prospect, it's important to remember that the NPA was still a small organization in 1973. It had just three employees: myself and Lesley Dunne, who were both full-time in the office, and my sister-in-law Mary Brennan, who was part-time. We had fifty-nine exhibitors at the 1972 Ploughing in Rockwell, County Tipperary, the year before I took over, compared with 1,700 nowadays. Back then, a site of 250 acres would have been more than adequate. Who would have thought we'd need a minimum site space of 650 or 700 acres some forty-five years later? After we hosted the World Ploughing Contest in 1973, there was a definite upswing in attendance and the exhibitors really became interested in what we were doing.

Initially I felt I was totally on my own, but then I realized the fantastic back-up I had with the executive committee. I recently found a photograph of that executive committee and I was shocked to find that I was the only one still alive. They were very staunch people and every one of them gave wonderful support to me.

We had set up a temporary office in White's Hotel, in Wexford, two months before I was appointed managing director, and I got great help from Dermot Jordan, who had a marketing role in Esso, our main sponsor at the time. They basically seconded him to work on the Ploughing with us. Esso was very good to us and, in fact, they promised to underwrite any losses that might be incurred from staging the world event, so that was very reassuring. I also relied heavily on the Wexford Committee of Agriculture – the committees of agriculture were the precursor to An Chomhairle Oiliúna Talmhaíochta (The Agricultural Training Council), known as ACOT, which was later replaced by the farm advisory body, Teagasc. Angela Cunningham from the

Wexford Committee of Agriculture carried out a lot of the local work.

The committee was based at Johnstown Castle and its head, Dr Aidan Conway, put all their services at our disposal for the event. We had meetings, functions and dinners at the castle and our foreign visitors were awestruck by it because of the beautiful setting. I also struck lucky with our site owners, Francis and Winifred Leigh, who provided Rosegarland Estate as the Ploughing site. They had a beautiful home and their door was always open to us. It was a perfect farm for the Ploughing because the fields were enormous – some of them were 70 or 80 acres. No matter who I turned to, before and during the event, people could not do enough for me and I got a real sense that everyone was wishing me well. That was a huge comfort, let me tell you.

A few years earlier we had thought it would be great if Oifig an Phoist, as it was then, could commemorate Ireland hosting the World Ploughing Contest with a stamp, so we wrote to them. It took a while, and several letters, but we were delighted when they finally agreed. They commissioned artist Patrick Scott to design two stamps. The 5p and the 7p stamp depicted a red-and-blue tractor ploughing, with a flock of seagulls following behind. As it happens, An Post, as it is now, is commemorating the National Ploughing Championships in 2017 with two stamps, which we are very happy about. We didn't have to beg this time around. We were amazed when they came to us, suggesting the stamp.

One of my first tasks as managing director was to prepare and agree a monument for the Cairn of Peace. I think this is a lovely idea. Every year, the country hosting the World Ploughing Contest erects a permanent monument marking the event. Every competing country brings a block of marble or stone, with an inscription in their native language for inclusion in the cairn. The motto of the World Ploughing Contest, *Pax Arva Colat* (Let peace cultivate the fields), is inscribed on the cairn. The finished work is unveiled on the first day of the contest and it is a beautiful and very emotional ceremony. The flags of the competing countries surround the cairn and the competitors stand behind their flags as they are raised during the opening ceremony. 'The Worldwide Ploughing Brotherhood', by Norman Appleyard, is played and there is a great feeling of solidarity between the nations. I love the sentiments in the song:

> From every country, race and clime
> We plough and sow 'til end of time
> United by the soil so good
> The Worldwide Ploughing Brotherhood.
> Firm is the hand that grasps the plough
> Warm is the heart of friendship now
> Remember as you turn each sod
> The Worldwide Ploughing Brotherhood.

I have been unable to track down where the song came from or to find out anything about the songwriter, but the wording suggests that it was written specifically for the World Ploughing Contest. The Cairn of Peace is seen as a symbol of the understanding, unity and friendship of people from all over the world. For the Wellington Bridge memorial we commissioned two sculptors, Patrick Roe and Philip O'Neill, to carve the granite monument, which stands on a base of Liscannor slate. It still stands on Rosegarland Estate and in fact a second one was added in 1981, when the farm became the first site to host the World Ploughing Contest twice. We also have cairns in Tullow and Oak Park in

Carlow, from other World Ploughing events, and they are all maintained by the county ploughing associations in co-operation with their county councils. Because of the dispute between JJ Bergin and the World Ploughing Organisation over the Northern Ireland competitors in 1954, no Cairn of Peace was erected in Killarney when it hosted the event. However, as I write, preparations are being made by a number of bodies, including the Kerry Ploughing Association and Kerry County Council, to remedy that, so we will have our cairn in Kerry someday soon.

Little were we to know in 1973 that the Cairn of Peace, and indeed the entire World Championships, would be completely overshadowed by a row over a race issue. Who would have ever thought that a ploughing contest could have found itself entangled in an apartheid controversy? Rhodesia, as Zimbabwe was then called, was still practising apartheid in 1973. We had understood that they were sending one white and one black competitor to represent the country, but it then emerged that they were sending two white competitors. All hell broke loose. The Irish Anti-Apartheid Movement (IAAM) called on the Government to ban them from entering the country. Kader Asmal, who was the public face of the IAAM, was very vocal about the participation of the Rhodesians and there was a suggestion that protesters might picket the grounds of the Ploughing. His name will ring a bell with many people as he was a great friend and supporter of Nelson Mandela. Professor Asmal taught law at Trinity College Dublin for twenty-seven years and co-founded the IAAM in 1963. He went back to South Africa in 1990 and was part of Nelson Mandela's team that negotiated the transition to democracy. He went on to serve as a minister in the new ANC government. Professor Asmal passed away in June 2011, but he'll always be remembered in Ireland because of his time in Dublin.

But back to the Rhodesians. They were travelling on British passports, because they both had British fathers, and therefore the Irish Government said it couldn't ban them. Then it was decided that the president, Erskine Childers, taoiseach Liam Cosgrave and tánaiste Brendan Corish would not be attending the Championships because of the controversy.

It was a tricky matter for the NPA and for the WPO, but either way we had no say in the matter as we were just the host country. Our own Association always stressed that it was non-political and non-sectarian and had no desire to get involved in politics. The rule from the WPO was that every affiliated society could send two competitors. Rhodesia wasn't breaking any rules and we didn't feel that we could step in and ban anyone from coming, but at the same time there was a lot of unhappiness about that country's participation. God knows, we didn't want to cause upset to anyone, but we were really caught in a bind. There were letters to newspapers complaining about the fact that we would be hosting the Rhodesians. People were trying to draw other rural organizations into the controversy and were calling on the Irish Farmers' Association and the Irish Countrywomen's Association to state their positions.

Michael T. Connolly was the Irish representative on the WPO and really he dealt with the whole controversy. I had no input at all into the discussion because my focus was on the National Championships, but it was an added worry. Luckily, we had a very strong executive committee, with several deep thinkers, and they offered Michael and me their full support. Michael went to meet Kader Asmal, to explain our predicament. Michael reasoned with him, stressing that we were non-sectarian and wanted to stay out of politics.

Michael was great at dealing with the press, and I remember all the journalists flocking around him asking him endless questions, but he was well able for them. His view was that people were entitled to express their opinion and to picket the Championships if they wished, but the organizers would continue to go about their work peacefully. He also criticized the Government for not sending a representative to the National Ploughing Championships, saying it was a separate event from the world contest and that the Irish organizers felt snubbed by the boycott. That was very true. We did feel annoyed that the Government was boycotting the national event, which had nothing to do with the world event. It was a regrettable situation, but what could we do? We just had to get on with things.

Michael's meeting with Kader Asmal must have helped because in the end the IAAM did not hold a picket at the event. They did, however, repeat their total opposition to the admission of the Rhodesian team. Kenya withdrew its two competitors on the eve of the contest in protest. And while we enjoyed fine weather for the opening of the National Ploughing Championships, a storm descended on the grounds in the days that followed and the heavens had well and truly opened by the time the World Ploughing Championships got underway. Nor did the Rhodesians have any luck. The champion and reserve champion prizes were claimed by Paavo Tuominen from Finland and our own John Traccy from Carlow, respectively. Despite the controversy, and the weather, we had a huge turnout over the four days.

It was a baptism of fire for me as managing director, and

something I could never have predicted when I took up the job. But to this day we remain steadfast in our view that we are non-political and non-sectarian and we leave politics to the politicians. Thankfully, we've never been involved in a similar predicament since. By the time we hosted the World Ploughing Contest again, in 1981, Rhodesia had become Zimbabwe. White-minority rule had been dismantled, and its prime minister was Robert Mugabe.

I still remember the enormous relief I felt when the event was over. It was like I had been holding my breath for the four days of the National and World Championships. I had such worries, inwardly, about the things that could go wrong: the possibility of an accident, a weather disaster, people leaving the Ploughing unhappy. Getting through such a big first event as managing director gave me great confidence for the next year.

My parents were delighted with my progress in the NPA, and my father could rightly take credit for getting me involved in the first place. Sadly, they didn't live long enough to see me rise to the top of the organization. I was almost thirty-eight when Mam died after a short illness, on 22 March 1972. She was seventy-six years old and her quick death was an awful loss. At least I had the consolation that my two children had had the chance to meet her and get to know her somewhat. She loved seeing the children coming, even when she was not so well. DJ and Anna Marie were three and five years old at the time and I remember she'd always have a bottle of Corcoran's red lemonade for them. She wouldn't let the adults have a drop of it! They were her only grandchildren at the time and she used to sit contentedly at the window for ages, watching them playing outside. DJ has many happy memories of the treats she used to give him. Anna Marie was too young to remember a lot of it, but she still remembers the Corcoran's red lemonade.

Her funeral made me appreciate how good we are in Ireland at handling death. I don't know what it's like in urban areas, but in the country the neighbours are so good when something like this happens. They arrive laden down with all kinds of sandwiches and cakes, and they are there to help with the livestock and the farming end of it because the family is taken up with funeral arrangements and meeting people. People arrived at my mother's funeral who I would never have expected to come.

For the first time, I appreciated how consoling it was to have a wake. It gives people a chance to mourn along with you and it does help in getting over the grief. If it's just you and the immediate family in the house, it's very sad. All the visitors give you a break from feeling lonely.

We have a tradition in our parish, if the bereaved family wants it, where a group of ladies get together and they provide tea and sandwiches in the local hall after the funeral ceremony. Nobody would refuse if they were asked to help out with it and it's a lovely, neighbourly gesture.

Mam's death was a huge break in the family because, as I said earlier, she was always there in the kitchen. It felt so wrong coming into the house and not seeing her there. There was an emptiness and a great loneliness in the house. She really was the centre of our lives.

Dad was very lonely after she died, but he was luckier than some because my brother JJ was living in the house with him. And of course, he had all of us around him like mother hens. He particularly knocked great sport out of DJ and Anna Marie. One time he brought them home two balls and they were delighted. But not long after, a fight got up between them. I don't know what Anna Marie did to DJ, but it resulted in a blood blister on his hand. In retaliation, DJ got her ball and kicked it into a monkey puzzle tree, bursting it. Anna Marie ran in screaming to Dad. On hearing both sides of the story, he said Anna Marie deserved it and that was the end of that.

Like many men of his age, Dad was one of those men who depended on their wives for everything, so he was quite lost when Mam was no longer there. Four years after Mam died, Dad had a massive heart attack and died in his own bed, on 29 October 1976. He was only seventy years old. His death was completely unexpected and came as a terrible shock to us. It's true that you grow up overnight when the last parent dies. As long as one parent is alive, you still feel like a child to some degree. You know they are always there if you have a problem or need advice. I always respected my parents' advice and had great belief in it. But after Dad passed away, I felt like I had to grow up overnight.

7. The Ploughing is all about the ploughing

I would say that about two-thirds of people who come to the National Ploughing Championships do not come to see the ploughing competitions. If the ploughing area is near the exhibition area, more people will go for a look, but many people at the event have no idea what's involved in the competition. For me, however, the most important thing about the Ploughing is the actual ploughing. Without it, we'd have nothing. And not all competitors are farmers, by the way. We have company directors, bank managers, bakers, nurses, teachers, all sorts of people taking part in the competitions. They've all been bitten by the ploughing bug.

People often ask me to explain how the competition works, and the best way I can do that is by equating it to cutting a batch loaf. So imagine I hand you a loaf and a knife and I give you a time limit and say that I want you to cut the loaf into slices. There is a catch, though: every slice has to be exactly the same size, shape, height and thickness. Now that's a different proposition. You would have to think very carefully, measure the loaf, calculate how many slices you could make and check with every slice that you were keeping to the right measurements, assessing each and every slice to ensure they were all uniform.

That's the way it is in ploughing, too: you give a competitor a very specific amount of ground – every competitor gets the same – and he or she has a number of specific things that must be done in that area. The competitors are judged on factors such as neatness, straightness, uniformity, packing and firmness. The perfect furrow is one that is well-turned and well-skimmed, which means the top layer is buried the full depth of the ploughing. Competitors must make sure that all the ground is turned over and that they haven't left any grass or stubble sticking up between the sods of earth underneath. The ploughed ground may look good, but it also needs to feel good under the judge's foot. It needs to be solid and fleshy.

In our ploughing contest, each competitor is given a plot of land, never any bigger than half an acre. They all line up and start at the same time – 10.30 am in our case. They do the opening split, which is the first turn of the sod of grass, to reveal the clay below. Then they stop while the judges examine their work. Straightness and uniformity are the key things to get right in the opening split. The plough must cut seven inches deep – the depth is measured by the judges. And they don't want to see any bits of stubble or grass showing in the upturned sod.

Once that part has been judged, the ploughmen and women continue to plough up and down their plots until the competition ends at 2.30 pm. At this point they must have their ploughs out on the headland – that's the strip of land left unploughed at the end of the field – or they are penalized. Even a few seconds over time causes penalties. You have no idea of the pressure when competing at world level because the time given to plough is shorter and the plots are bigger. I'm always a nervous wreck watching the Irish competitors. Some of them really cut it fine, timewise, and when I say fine, we are talking about driving out of the plot with seconds to spare. Eamonn Tracey is a prime example. He will surely give someone a heart attack yet – most likely me.

But he just jumps off the tractor and says: 'Sure I knew I had five seconds to go.'

I was the timekeeper when Martin Kehoe was ploughing at the World Ploughing Contest in Kenya in 1995, and I remember watching him and thinking he was going to run out of time. I was certain that Martin was about to lose a world title because of me signalling the final whistle. I returned to the Irish team crestfallen, but they were in very high spirits. I don't know how, but Martin had pulled out of the plot with seconds to spare and he won the gold medal.

The focus of the competitors is unbelievable when they are preparing to start ploughing. It reminds me of a surgeon getting ready to operate when you see them laying out their spanners, setting their ploughs and having the plough boards shining like a silver platter. If there's one missing link, the whole operation could go wrong because timing is so crucial. A missing wrench or measuring tape could really throw the competitor off his game. But in fairness to them, the competitors always help each other out if someone has lost an item or forgotten something.

I often think ploughing is like baking a cake. Sometimes it comes out beautifully and other times it just doesn't go your way and there's nothing you can do. I have to put my hands up, though, and say that I've never ploughed a field in my life. I don't know why I didn't give ploughing a go when I was younger. And yet my sister Eileen did. But Eileen always loved a challenge and she would try her hand at anything, and indeed excel at it.

While I never ploughed, I can tell everyone how to do it, and that's from years of looking and listening and clerking to judges. You learn what the judges are looking for and you know what good ploughing looks like. I would never volunteer to judge but if someone was stuck, I would be able to do it. It's quite a skill to become a good judge. As in most sports, the judge or the referee is not always considered to have the right opinion, so we have a team of three or four on each set of plots. We deem their decision to be final, so no questions are permitted after they have judged.

I'd recommend anyone to have a look at the competitions the next time they go to the Ploughing. Even if you know nothing about ploughing, you would see the difference between a very well-ploughed plot and a middling one. If it does not look good and straight and the sods don't look similar and you can see lots of grass or stubble sticking up between the sods of ground, then the competitor will be penalized.

So that's how the ploughing competition works, but I should also explain that there is more than one type of plough. Our competitions involve conventional and reversible ploughs. Conventional ploughs can be two-furrow or three-furrow — that is two boards or three boards on the plough, the boards being the parts that cut into the ground.

A reversible plough has two or three boards on the bottom of the plough and the same on the top of the plough. This is quicker to plough with because one set of boards ploughs to the left and one to the right. So when the competitor gets to the end of the plot, he just turns over the plough and goes back in the direction he has come, instead of going around in big loops. It would be a matter of opinion as to whether conventional or reversible ploughing competitions are harder to win, but the two titles have equal merit at world level.

And then we have match ploughs and standard ploughs. The difference is similar to the difference between a racing bike and an ordinary bike.

We have more than twenty competitions and various age categories at national level, including the supreme conventional and reversible classes, which select the two competitors for the world contest. The World Ploughing Contest keeps it simple – they just have the conventional and reversible classes, on grassland and stubble ground. Some thirty-three countries are now affiliated to the WPO and last year fifty-eight competitors took part in the World Ploughing Contest at York, England. Thirty competed to be world champion in reversible ploughing, while twenty-eight took part in the conventional contest. Every competitor has to start at club level every single year, and they have to qualify from their county to compete at national level – no matter if they were world champion the previous year.

We also have vintage ploughing classes, where tractors must have been manufactured before 1959 and match ploughs are not allowed. We have several horse classes and we are determined to keep the art of horse ploughing alive because it's how we started. This year we introduced a special grant for each county, to encourage them to bring along a new horse ploughing competitor. The teams of horses are fabulous to watch and you feel like you are rolling back the years when you watch the horse and his master, or mistress, working peacefully together.

We are rolling back the years even further with the loy digging competitions. The loy is a narrow spade with a foot-rest, which was used to turn the sod before horses and ploughs and tractors took over.

All in all, we could have about 350 competitors taking part in the National Championships. What a parade that is, and I am at my proudest when I go to the marshalling yard and see the competitors preparing to compete for an All-Ireland title. Some are dreaming of that first win, while others are going for their fifth or tenth title.

It really grounds me when I get out of the craziness of the exhibition arena and just take stock of the competition fields and see how important the actual ploughing is to thousands of people. Ploughing has a language of its own and no matter where we are in the world, if there are a few ploughmen or women in the group, the conversation will sound something like this: 'His ins and outs were bad' . . . 'his middle was too high' . . . 'his furrow was too crooked'. I am always amazed at how the competitors can remember exactly the reason why they won or lost a particular match some twenty or thirty years ago.

I really love going to our awards night, which takes place a couple of weeks after the National Ploughing Championships. We have more than 600 competitors and officials attending. No dignitaries, no media, just pure ploughing folk, and that's what makes it special. We used to have it on the last day of the Championships, but everyone was so exhausted after the Ploughing that people were in danger of falling asleep into their soup, so we pushed it on a few weeks.

We have plenty of political dynasties in this country, but they don't have a patch on our ploughing dynasties. Four generations of the Traceys from Knocklonagad in Carlow have left their mark on the plough. I first met Mick Tracey through the Ploughing, and a fine ploughman he was. Then his son, John, took silver in the World Ploughing an incredible six times between 1973 and 2009. John's son, Eamonn, took over and was runner-up in 2012, world champion in 2014 and 2015 and runner-up in 2016. And now Eamonn's son, Seán, is a very promising ploughman. He has been an under-21 national champion several times over and I have no doubt we'll see him on the World Ploughing stage before too long.

Wexford is famous for its ploughmen, so Martin Kehoe from Foulksmills took it all in his stride when he was crowned world champion three times, in 1994, 1995 and 1999. He took silver and bronze in 1992 and 1993, respectively. And why wouldn't he? His father, Willie, won the first of many national ploughing titles in 1941, back in the pre-tractor era when all the ploughing was done by horses. And then in 1955, at our silver jubilee in Athy, Willie won the national title with tractor ploughing and thus made history by becoming the first person to win national titles in both tractor and horse ploughing. That feat has not been repeated since.

And now another generation is taking up the challenge. Martin's sons, Willie John and Martin Jr, have won their own national titles. Willie John has represented Ireland at three World Ploughing Contests. Their sisters, Eleanor, Christine and Michelle, are fine plough women in their own right and every one of them has been crowned Queen of the Plough. Isn't that some record for one family? And Martin has several other strings to his bow – he is a world champion in the discipline of tug-o'-war and he also found success in activities such as weightlifting and sheaf-throwing. The man can turn his hand to anything.

John Whelan from Ballycullane is another one of Wexford's ploughing giants, despite the fact that there was no tradition of ploughing in his family. He won the world championship in 2013, came third in 2014, second in 2015 and third in 2016. His son, Stephen, is now competing and doing the family proud. There are so many ploughing dynasties from around the country that I would surely forget one if I tried to name them all, but I can't leave the subject without name-checking the two generations of Keatings, Coakleys and O'Driscolls who have also ploughed at world level.

That's not bad going for a country the size of Ireland. When you see the training that these people put in to qualify for the World Championships, you're left in no doubt that competitive ploughing is as much a sport as soccer or rugby. In fact, we have approached the National Sports Council on several occasions to have ploughing recognized as a sport, but we've made no progress. We've made it clear that we don't want funding, in case that's the sticking point preventing them from recognizing it as a sport. We just want to see our competitors getting fair recognition for their achievements and we think that they will get that kudos if they are seen as sportsmen and women. Who knows? Perhaps, some day, the National Sports Council might change their minds.

Austria would be one of the strongest countries in the world ploughing stakes. They take it very seriously and fair play to them. When their team is selected for the World Ploughing they get special training before the competition and they go to the host country earlier than everyone else to get acclimatized. They secure land away from the site to practise on for weeks in advance, so that they get used to the soil and conditions. We also take land in the host country to practise on, in advance, but no one prepares as rigorously as the Austrians.

The Scottish competitors have also been doing very well in recent times. Last year, a father and son, Andrew Mitchell Snr and Andrew Mitchell Jnr, won the world titles for conventional and reversible ploughing. That had never been achieved before and it was fantastic to see them doing so well.

There's no doubt that ploughing people are in a league of

their own. Take the late David O'Connor, a Wexford plough-man from Cushinstown who was a very dear friend of mine. He was competing in the 1938 Championships in Oak Park, County Carlow, and here's how he got to the event: he took a bus from New Ross to Tullow, then he walked seven miles to the site, wearing a new pair of boots. He told me he had to take off the boots halfway because of the massive blisters on his heels. After all that, he was introduced to the horses he would plough with and he had to contend with a plough he'd never seen before. And despite all that, he won the national title. I don't remember how he got home, but sure wouldn't you skip home after winning?

David was on the NPA's executive committee when I took over as managing director and we became great friends. Ploughing people are so dedicated to their work. I remember visiting the site in Ardfert, County Kerry, a while before the Ploughing in 1984 and I met Louth competitors coming out of the farm. Can you believe they were after travelling all that way to take a soil sample so that they would know the type of land they'd be ploughing? They'd want to know if it's going to be sandy soil or heavy soil and prepare themselves for that. That gives you an idea of the dedication of some ploughmen.

When the Ploughing was in Ballacolla, County Laois, in 1995, local farmer Robin Talbot provided stabling for the ploughing horses. I remember him telling me how he would be woken at dawn by the clip-clop of the horses being brought out of the stables for exercise. At dawn!

We had a lovely moment at the 2014 National Ploughing Championships in Ratheniska when Charlie Keegan's grandson, Michael, arrived with the tractor on which Charlie had become Ireland's first world champion in 1964. Michael had spent nearly a year restoring the 1964 Deutz D40L tractor after tracking it down to a farm in Kilcoole, County Wicklow. He also secured the exact model of the plough that won the competition. His grandfather had used most of the original plough – a 1964 Kverneland Hydrein match special – for spare parts, but Michael had managed to incorporate the remaining pieces of the original plough into the plough he tracked down in England.

There's something poetic about ploughing, so it's not surprising that it has inspired many ballads and poems. There are odes to ploughing champions, such as Wexford's Michael Redmond, who was champion ploughman for more than twenty years, from 1932. The author's name has been lost, but the ballad goes like this:

> Nine times I saw Mick Redmond
> Take honours that were high
> From Slaney's banks he came and turned
> Nine counties' soils to sky.

Dan Holland penned a lament for Cork's Jerry Horgan, who was a wonderful horse ploughman:

> How pleasant sure to watch him
> And how he would guide the plough
> The big wheel in the furrow
> Well in against the brow.

Jerry is also remembered by a memorial in his native Ballinagree, which sits beside a memorial to another famous Ballinagree man – the bould Thady Quill. Cork has a huge tradition of ploughing, and competitors would often have to travel over 50 miles by tractor just to take part in a qualification match.

I was glad to see the many talents of our champion

ploughman Martin Kehoe being brought together in a song by Pat Fortune:

> The lea or the stubble
> To him was no trouble
> Perfection in all he had done
> He went off to Dunedin an All-Ireland champ
> And returned as the world's number one.

I mentioned some of our founder JJ Bergin's achievements earlier on, but I didn't include his talents as a writer. JJ wrote 'The Song of the Plough', to be sung to the tune of 'The West's Awake':

> Turn down the green, O man who ploughs:
> Guide thou the plough with sharpened share!
> Turn up the brown to sapphire skies!
> Mankind on thee for bread relies.
> Bright shines the sun and God looks down
> On man, on beast, on hill and town
> Then sow the seed in mellowed earth
> To harrows away and wild birds' mirth.
>
> The joyful hum of threshing time,
> And later drone as mills make flour,
> Mankind gets bread: but what man thinks
> It was your sweat that forged the links?
> But, sure, the world must bend its will
> In every age to ploughman's skill:
> Then, Oh! Hurrah, all men who toil,
> You're masters of the sullen soil.
>
> Turn up the brown, O man who ploughs!
> The waken'd earth to warming sun,

And give all men their daily bread,
Your work is God's for He has said
He'll bless your work, your plough-team too,
Reward is sure for what you do.
Then, Oh! Hurrah, sons of the soil,
God speed the plough, God bless your toil.

I'm not sure when it was penned, but it first appeared in a National Ploughing Championships catalogue in 1942. Underneath the song were a few lines which dictated that: *'No social function in rural Ireland should be concluded without this song. It should be included in the programme of every ceilidh, country dance and concert. All singers who are proud of their connection with the soil of Ireland should add The Song of the Plough to their repertoire.'*

I'm guessing JJ Bergin wrote those lines in the hope that his ballad would survive and prosper. And indeed the song is still included in the Ploughing catalogue, some seventy-five years after that recommendation was made. Other song-writers and musicians to celebrate the Ploughing include PJ Murrihy and Richie Kavanagh, while composer and musician Ollie Hennessy wrote a special ballad for the 2006 World Ploughing Contest.

Ireland does very well on the world ploughing stage. Our top ploughmen often have to deal with soil conditions that would not be normal here in Ireland. The competitor who has the wit and energy to adjust and adapt correctly on the move will always perform well.

I don't see how the standard could go much higher, but it probably will. I mentioned Charlie Keegan earlier – the first person from the Republic of Ireland to take gold in the World Ploughing Contest, in 1964. There's a plaque and a granite bench commemorating his achievement in his home town of Enniskerry, in County Wicklow. Charlie was one of the first people to bring a world title of any type back to Ireland, and it was a massive thing for the whole country back then. I suppose it would be like Ireland winning the World Cup now. When he arrived back from the contest in Austria he was carried shoulder-high off the aircraft with the Golden Plough trophy in his arms. The Artane Boys' Band were out in force to welcome him home. An open-top bus, followed by a motorcade, took him to Enniskerry, and bonfires were lit along the way to congratulate him.

That warmth towards ploughing we see in Ireland just doesn't exist in other countries. I know that the World Ploughing competitors love to come here when we host the contest because the Ploughing is so highly regarded here.

Since 1960, I have travelled to seventeen countries for the World Ploughing Contest, some on more than one occasion, and every time I've travelled I've felt so proud of our Irish competitors. They are such good ambassadors for Ireland and the comradeship among ploughing enthusiasts is very, very special. I suppose travelling the world together brings people closer, too. A world team could be up to three weeks together in very close proximity, so you have to get along. It's no different from the Irish team going to the Olympics. As well as sending two competitors, our team includes a coach/team manager and a judge. Every participating country sends a judge to the World Ploughing Contest.

Getting our competitors and their equipment to the World Ploughing Contest takes a lot of planning. The tractors and ploughs go off on a container weeks before the event. If the event is outside the European Union, we have to prepare a legal document, called a carnet, which we lodge in Dublin

Chamber of Commerce. That document itemizes and prices every item in the container, from a spanner to a plough part. Many ploughing societies abroad cannot afford to send their competitors' tractors and ploughs to the world event, but we make a point of doing it because we believe it makes a huge difference to the team. The host country organizes the loan of tractors and ploughs for the event if competitors do not bring their own, but it really puts them at a disadvantage if they have to familiarize themselves with a different tractor and plough just days before the event. Some countries send only their ploughs, but there can be endless problems with tyre size and attachments. So we think it's worth the expense.

This year, in 2017, the World Ploughing Contest takes place in Nakuru, Kenya, in December. The tractors and ploughs will have to leave on the evening our National Ploughing Championships finish because it could take the container between ten and twelve weeks to arrive at the site. A day late, and a competitor could miss his or her chance of a world title. It will cost us about €17,000 to transport the equipment to Kenya.

Hosting the World Ploughing Contest is a costly exercise, but it gives a great boost to the ploughing sector. It cost us a lot of money to stage the 2006 World Ploughing Contest in Tullow, but we felt it was money well spent. The host country looks after the costs of four team members from each participating country for the duration of the ten-day programme. Items covered include accommodation, meals, transport, tours, entertainment, mechanical back-up and many other expenses. The team includes the two competitors, the WPO board member, the judge and the team manager. Many competitors bring a support team with them, but the host country is only obliged to look after four team members.

I love travelling abroad and meeting colleagues from other countries. Of course, some of them have preconceptions about Irish people, and it is funny to see the reaction of people when you tell them you don't drink. They are always taken aback. 'You're Irish and you don't drink?' they say, shocked. We're so well known for the drink.

I remember going to the World Ploughing in the Netherlands and this lovely farmer was hosting me and a colleague on his farm. Back then, the travelling teams were accommodated by farmers, but nowadays we stay in college campuses or hotels. Anyway, at the end of the day we were having a meal and the farmer offered us wine. I explained that I didn't drink and he asked if I'd take a glass of milk. Thinking that it would be a nice cool glass of milk from the fridge, I said: 'Yes, please.' But then he disappeared outside and came back a few minutes later with a glass of warm milk, straight from the cow. Well, my stomach was churning but I couldn't be rude, so I had to drink it. I thought I'd never see the bottom of the glass.

There's one thing I hate, and that's being in a room far away from everyone else when I'm in a hotel abroad. I remember being put in a room on my own at a World Ploughing Contest in New Zealand. We were staying in this big college and I was at the far end of it while the other people I was travelling with were at the other end.

Anyway, I was sitting at a board meeting and some movement outside the window caught my eye. Two of the women I was travelling with were carrying this big mattress past the window. They were moving my bed into their room. I pretended I didn't see them, but I had to put my head down before I started laughing. There is such camaraderie among us when we travel, and sure, if you don't have that bit of fun, what's it all about?

I've made no secret of the fact that I'm a home bird, and I was quite nervous staying away from home in the early days. I remember one occasion when I was staying in a room by myself in Paris when we went to the World Ploughing in 1961. One of the competitors' wives – Mary Teresa Murphy – offered to share the room with me when she heard that I was a bit anxious. And I remember we got this big heavy chair and put it under the knob of the bedroom door so that we would hear the commotion if anyone tried to break into the room. I wouldn't mind, but there would have been a lock on the door. As if anyone would be bothered breaking in to the likes of us!

On the same trip, we went to that cabaret music hall Folies Bergère in Paris. Lots of high-kicking dancers and if you put all their clothing together, it would just about decently cover one person. One of our directors, who would later become our chairman, Larry Sexton from Courtmacsherry in Cork, was with us. He was a very good-living, strong Catholic man and he had two sons who were priests. And when we went into the show, all I could hear was Larry saying: 'Say your prayers, Anna May, say your rosary.' Sure, it was innocent enough when you look back on it now. The same Larry would always insist on walking on the inside of the footpath when we were abroad, to make sure that I didn't wander into a shop. He knew the measure of me and was ready to halt my gallop if I headed for a shop entrance!

The Ploughing has provided so many memorable moments in my life – some of them comic. Like the time we were at the World Ploughing Contest in Lincolnshire in 1984. It was very mucky underfoot and getting around was quite diffi-cult. I well remember warning some of our supporters to be careful of the marquee guy ropes – those ropes that tie the tent to the ground. Anyway, I mustn't have been listening to my own advice because when I was rushing around didn't I become entangled in a guy rope and I fell flat on my face in the muck. I was covered in mud from head to toe. I was due to meet some dignitaries, but I can tell you I was in no fit state to meet any dignitary after that. A nearby exhibitor offered me his washing facilities so I was able to make myself a bit more presentable. I heard that one of our supporters said: 'Anna May has fallen in the muck. We may get a JCB to lift her up.' The devil!

A memorable moment that was very moving occurred at the World Ploughing in Masai Mara national park in Kenya in November 1995. We had huge language barriers with some of the countries and it could be hard to get a conversation going. One evening we were just after having dinner and everyone was sitting around. A Slovenian delegate, Alojz Avzic, had a concertina and he started playing something quietly. Gradually we realized it was 'Silent Night'. They started singing it in their language and we joined in, singing in English. It was spine-tingling. We couldn't say a word to each other because of the language difference, but we could sing together. I will never forget that lovely memory and I can still picture that man sitting there with the concertina on his lap.

Music is great for uniting people. I think it's so important that children are taught music in school. There's so much time given over to sport, and rightly so, but music is so important and you'll have it with you for life. (My musical favourites are singers like Josef Locke, Finbar Wright, Finbar Furey, Andrea Bocelli and, of course, our own Daniel O'Donnell. I was very fond of Pavarotti, but then the old devil left his wife after thirty-four years of marriage and married a much younger lady. He disappointed me with that behaviour and I fell out with him after that.) I've met the country music singer Derek Ryan at the Ploughing and found him to be an absolute gentleman, so I'm delighted to see him doing so well.

On that same visit to Kenya, I remember going to a market in Nakuru and meeting this woman who had a clutch of chickens in a box. She was sitting on an old block of timber and singing her heart out. She looked so happy that I just had to talk to her. She smiled at me, looked up at the sky and said: 'I'm singing to God.' She was so happy, despite all the poverty around her. I was really struck by that and I've never forgotten her. We complain about so much, yet we have incredible wealth when we compare our lives to the life of that woman in Kenya.

In the early days, I was like the mother hen at the airport because the competitors wouldn't move a step without me. In fact, they'd come to my house and we'd leave for the airport together. I suppose people wouldn't have been used to air travel back then.

We had an eventful dash to the airport when we were going to Zimbabwe in 1983. Competitor Willie Ryan from Kilkenny had to stop for a part for his plough at Larkin's, in Baldonnel, on the way. We were running very late at this stage and when we got to the airport I discovered that Willie was checking in approximately half a plough with his luggage. I can still see him leaving the board of the plough on the conveyor belt at the check-in desk. There were bits everywhere and I thought to myself: 'Half of that will never be seen again.' David O'Connor was our WPO board member that year and he had labelled everything neatly and left nothing to chance. Guess whose luggage got lost? Yes, David's never turned up, while every bit of Willie's plough came rolling out on the baggage belt in Harare.

Of all the countries I've visited, I suppose New Zealand is the one country I felt a very strong connection with. I'm very fond of the New Zealanders because they are so like the Irish, and indeed there are so many Irish living over there. We were there for the World Ploughing Contest in 1980 and at one particular gathering, someone in our group sang 'Danny Boy'. I saw the tears running down the faces of some emigrants. Even though they had made their homes in New Zealand many years before, they still felt homesick when they met their fellow countrymen and heard the songs from home. It revived the Irish in them. I remember the last time we visited and we were saying goodbye to our New Zealand friends and, I suppose, silently wondering would we ever meet again. We were shaking hands and we were so reluctant to let go that we were holding on to each other's fingertips in the end.

We've made so many friends abroad because of the Ploughing, and if John was alive he'd tell you about the number of times that people turned up on our doorstep because I, or Anna Marie, had invited them to visit us. Of course, when they'd arrive we wouldn't be at home at all, and John would be left to entertain them and make small talk until we got back. One time a New Zealander arrived on the doorstep after attending the World Ploughing here. We were finishing up at the Ploughing, so we told him to go ahead to the house and we'd follow him. John was just back himself and of course there wasn't a thing to eat because we'd been gone for the week. He was mortified because there wasn't even milk for the tea. He told the man he had to run down to the shop. He got back, made the tea and poured for the visitor. Then he offered him the jug of milk. 'I don't take milk, thanks,' the man said. How we laughed when John told us that story.

That sort of thing happened regularly. 'Bad scrant to ye women,' he'd say when we'd get back. That was the nearest thing to a curse John would ever utter because he wasn't a man to use swear words. 'Bad scrant to ye,' and he'd shake his head, and we'd all laugh at his mock annoyance, John included.

8. How the Ploughing just grew . . . and grew

The film *The Song of Bernadette*, based on Franz Werfel's book of the same name, has a quotation at the beginning: 'For those who believe in God, no explanation is necessary; for those who do not believe in God, no explanation is possible.' Some people attribute the quote to St Ignatius of Loyola, while others believe St Thomas Aquinas said it first. Regardless of the source, it is a quotation that has often been used by journalists in reference to the Ploughing. They have tweaked it slightly to say: 'For those who attend, no explanation is necessary; for those who don't, explanation is impossible.' And it's very apt because if you've never been to the Ploughing, it's hard to get your head around the scale of it all.

It's a ploughing match, yes. But it's a machinery and live-stock exhibition. It's a great meeting place. It's a shop window for crafts and local enterprises. It's a fashion show, a motor show and a feast of food. It's a mini town where you can buy everything from a hurley to a combine-harvester. There's a hunt chase, there's sheep shearing, horseshoe-throwing and vintage machinery. I could go on and on, but I might bore you. It's really a showcase of Irish food, farming and culture. We had almost 1,700 exhibitors in 2016 so if you were to spend just four minutes at every stand, it would take almost five days working around the clock to visit each stand, not counting the time it would take to get from one stand to the next.

It's the biggest outdoor event in Europe and it's the biggest ploughing championships in the world. I know that's true because I've travelled halfway around the globe to World Ploughing events. For some reason, other countries have not expanded in the same way as we have. Most of them have opted to concentrate on the ploughing, which has limited their growth because they don't get the income from the exhibitors and the visitors. And they don't attract massive numbers of non-farming visitors because they don't have the huge array of attractions that we have. You'll see everything from Friesian cows to food demonstrations at the Ploughing and I think that variety is why people like it so much.

'Keep calm and carry on ploughing' would be a good motto for us when you consider the various challenges we've met to get to where we are now. Every year we see an increase in attendance, and it has really soared in recent years. It went to a two-day event in 1955 and increased to three days in 1988. Last year we had 283,000 in attendance at Screggan, County Offaly – an all-time high. In 2017 we expect as many again. Attendance figures are scarce for our earlier events, but the first-day attendance in 1972, just before I took over as managing director, was 10,000 people, so the total attendance hardly topped 20,000 over two days.

The Ploughing means an awful lot to the economy when you think of the money people spend on accommodation, food and petrol. And that's before they buy anything from the exhibitors. The first thing we do before we announce where the following year's Ploughing Championships will be held is to block-book rooms in the nearest big hotel for all the key workers at the event. Otherwise they'd be booked out within hours, once the venue is announced.

In 2011 we asked University College Dublin to carry out a

survey on the economic impact of the Ploughing. We always knew it generated a huge amount of money for the local hotels and businesses and for the exhibitors, but we thought it would be interesting to put a figure on it. The researchers reckoned it was worth €36 million to the economy at that time. That figure has certainly increased since then, considering we were only starting to emerge from the recession at that time. It would be interesting to repeat the survey now, given the upswing in the economy. If the harvest is good, the farmers are all out and willing to spend, and the atmosphere is fantastic.

We never had a masterplan to grow the Ploughing into what it is today. It all grew naturally. Take the livestock section. That came about by accident. David Hutton-Bury was hosting the 1987 Ploughing on his Charleville Estate in Tullamore. We were inspecting the site months beforehand when one of the local organizers, Seán Healion, pointed to the farm sheds and said, 'Why don't we have a few cattle in the sheds at the Ploughing?' It was a great idea, and now all the cattle and sheep breed societies exhibit. It's wonderful to see the best of our livestock all gathered in one place.

Similarly, the very popular hunt chase came about because David Lalor was hosting the event on his farm in Ballacolla in 1995. He is a very keen huntsman and he got us interested in the idea of a hunt chase, so we decided to introduce that.

And then there was the fashion show. Let me tell you, there were a few eyebrows raised when I came up with that idea! It was 1984 and we were meeting in the Killeshin Hotel in Portlaoise to discuss the upcoming National Ploughing Championships in Ardfert, County Kerry. When Any Other Business came up, I said to the men that I'd like to have a fashion show. They all looked at me and this man stood up and said: 'Mr Chairman, I thought it was a ploughing match we were running. What does fashion have to do with ploughing?' End of story. There was no discussion and not another word said about the fashion show.

I went out of the meeting and said to myself that we had to get something for the ladies. I mean, if you're not interested in farm machinery or ploughing or livestock, what would bring you to the Ploughing? So, we went ahead with the fashion show. And I remember trying to get into the tent when the first show was on and it was absolutely packed. There were three men standing at the doorway and I couldn't get by them. And one of them was on a walking stick. What were they glued to, on the catwalk? Wedding gear! I said: 'Excuse me, can I get in here, please?' and they said, 'Well, we're not moving.' It was comical. I had to go around to the back of the tent and go through the dressing room to get in.

We just had one fashion show on one day of the Ploughing at the start, but the demand was so great that we now run several shows every day. And while it was my intention to give women something to look at, the men are definitely as keen as the women when it comes to the fashion tent. There was never another word said about the fashion show after the first year. And they freely admit now that it was the right thing to do.

We have a lot more to offer non-farmers these days. I'm thinking of the shopping arcades, craft and enterprise villages, cookery demonstrations, innovation arenas, education hubs – all sorts of attractions to ensure that there's something to interest everyone.

We try to introduce new things all the time to keep people's interest. We had the Dancing Diggers doing great stunts in 1996 in Carlow. They came from America and it cost quite a lot to bring them, but they were a big hit with people. And of course there was the pole-climbing in Athy and Ratheniska, where competitors chased up poles more than 30 metres tall, as fast as possible. The speed of them was incredible. We also had The Axe Factor, where trained axe men did speed and skill trials with hatchets, and of course welly-throwing is a mandatory activity at the Ploughing.

The robotic milking machines drew great crowds in recent years. The interest in that was amazing. People would stand for hours watching as the cows queued up to be milked. The gates would open automatically to let them in and the cows would stand patiently as a robotic arm cleaned their udders and put on the milking clusters. The electronic collars around their necks were recording all their movements and milk yields, and if an animal should not be milked for some reason, the gate would not open when it identified the cow. Isn't that a long way from the three-legged milking stool?

In 2016 we had this crazy tractor football where tractors were driving after a giant ball. It was quite a novelty and people seemed to enjoy it.

With more than 280,000 visitors, the Ploughing is definitely a fertile ground for spreading your message. We've had all sorts of religious groups setting up stands with us. The only one we refused was the Church of Scientology. We just felt that we couldn't facilitate them. The man who had tried to book the place was very annoyed when we turned him away. But people forget that we do have the right to reject any application for exhibition space and we don't have to give a reason. It's the same thing if we get too many people trying to sell things like sunglasses and batteries. We have to ensure that we have the right proportion of exhibitors in each category and we are very careful not to saturate the retail space. We could easily double our takings in this area by letting everyone in but that's not good for the visitors, or indeed for the retailers.

Sometimes people might get on to their local TD if they're not happy that they've been refused. I remember one such case when we turned down an application for something similar to the sunglass stand. The woman said she was getting on to her public representative. And who was that but Dr Ian Paisley, the former DUP leader who has since died? The Lord have mercy on him. Anyway, he rang me up to talk about it and I was very nice to him. I said: 'Look Mr Paisley, if you do your business, I won't interfere, and you do likewise with me.' I wasn't afraid to say it. I treat everyone the same. And that was that. He never came back to me. I met him afterwards at one of the Northern Ireland ploughing matches and he was very nice. He was a great supporter when it came to agricultural events and I know he was very highly thought of by farmers in Northern Ireland.

The recession certainly left its mark on our machinery and motoring exhibitors. I remember we went from a massive car display to just about four jeeps in 2010. Now all the motoring people are back and firing on all cylinders again.

There are always a few stands that people are determined to visit. Would you believe that one of the most popular stands at the Ploughing every year is Met Éireann's? People love to meet the weather people from the television and see how the forecast is put together. They have all their digital technology there to show how they can predict what's on the way – very important if you are thinking of making hay.

The people who attend the Ploughing are also in a league of their own when it comes to enthusiasm. The official opening time of the event isn't until 9.00 am. We arrive on the site at 6.00 am to get organized and we're always amazed to see the first visitors starting to arrive half an hour later. They might have been travelling from 4.00 am to get there that early. I remember years ago people wouldn't arrive until 11.00 am, but there wouldn't have been that much to see back then. These days, it's often the case that a day isn't enough to see everything. It's quite normal at every Ploughing that people would call into the NPA office to ask if there was anywhere to stay locally. They would have decided to stay another day because they hadn't seen enough. Now you wouldn't have a hope of getting anywhere local, but we'd always find somewhere for them.

RTÉ puts on a special programme every evening from the Ploughing and that is really enjoyed by the people who cannot get there for whatever reason. Older people who can no longer go to the event tell me they truly appreciate the fact that they can watch it on television each evening. And it's good for people who can't move around, maybe because of bad hips or whatever. They get a real taste for what's going on, between the farming and the social side. I think it's a true public service. It also shows how important the event is. There's no way RTÉ would have invested so much in the Ploughing fifteen or twenty years ago.

I already mentioned some poetry and ballads that have been inspired by the Ploughing. Now I have to mention one that I'm particularly fond of. If I tell you it's called 'The Ballad of Anna May' you might understand why! It grew out of a late night gathering of a group of journalists in a Chinese restaurant in Portlaoise. Larry Sheedy, the man who compiled our sixty-five-year history, was there with the group, which had gathered after a long day at the Ploughing in Ballacolla in 1995. It included Willie Dillon, who was then the *Irish Independent*'s agriculture correspondent, IFA press

officer Niall Ó Muilleoir, *Irish Press* staff and Seán MacConnell, who was *The Irish Times*' agriculture correspondent.

Seán, who sadly passed away unexpectedly just before the
Ploughing in 2013, was a great raconteur and singer and came
from a very musical family. His brother Mickey wrote 'Only Our
Rivers Run Free', among other songs, while his brother Cormac
wrote 'A Silent Night – Christmas 1915' about one of the Christmas truces during the First World War. Anyway, the assembled
group decided Seán would be best placed to write a ballad about
the National Ploughing Championships. According to Larry, it
took about an hour, and a lot of paper napkins, but with the help
of the scribes, Seán produced 'The Ballad of Anna May'. I know
it's very complimentary to me, but it also captures the scale of
the event. It's also very funny and it's a lovely memory to have of
Seán, who was a wonderful man to know.

> Oh England has its Princess Di
> And France Brigitte Bardot
> And there are lots of lovely ladies
> Any place you have to go
> But beauty is a passing trait
> And you cannot it renew
> So here's a toast to Ireland's best
> She's Anna May McHugh
>
> Helen's face it launched a thousand ships
> And Gráinne Mhaol was bold
> And there's tales of Annie Murphy
> Which can never more be told
> But there's no trace of scandal
> They're not fit to shine her shoe
> For she launched a thousand ploughmen
> Our Anna May McHugh

My parents, James and Elizabeth Brennan (née Wall), on their wedding day in 1928.

My mother with Betty (*standing behind her*) and (*from back*): me, baby Oliver, Stannie (*front*); JJ, Gerald, Paddy and Shep the dog, in 1938. Dad must have taken the photograph.

At the Royal Marine Hotel, in Dún Laoghaire, wearing the dress I had made for my twenty-first birthday in 1955.

The 1959 Ballylinan camogie team. I am third from the right at the back and my sister Eileen is first on the left, also at the back.

With my husband-to-be, John McHugh, at Jervis Street Hospital Nurses' Dance at the Gresham Hotel, Dublin, in the 1950s. My sister Betty was a nurse there.

My sister Eileen, leading the St Patrick's Day parade through Dublin in 1962, as was the custom for the woman who won the farmerette class at the previous year's Ploughing.

Eileen with her Queen of the Plough crown and cup, after she won it for the first time in 1961, with my mother and father. They are standing in front of our home place in Clonpierce.

Our first photograph as husband and wife: leaving the Sacred Heart Church in Arles on our wedding day, 28 July 1966. Our best man, behind us, was John's cousin, also John McHugh.

With my family on my wedding day: (*left to right*) Oliver, Stannie, Eileen, Betty, John McHugh, me, my father James, mother Elizabeth, JJ, Paddy and Gerald.

With one-month-old Anna Marie, two-year-old DJ and John, in 1969.

With Anna Marie, aged four, at the World Ploughing Contest in Wellington Bridge, County Wexford, in 1973.

John with Anna Marie and DJ on their Confirmation Day in 1980. Note the absence of Mammy – I was at the World Ploughing Contest in New Zealand.

Pictured with my family, John's sister, Sister Dominic, and NPA and ICA members when I received the *Leinster Express*'s Laois Person of the Year Award in 1991. (*Leinster Express*)

DJ, John and I celebrating Anna Marie's conferral with a Diploma in Marketing from the Institute of Technology in Carlow, in 1989.

With DJ, his wife Fiona, Anna Marie and her husband Declan when Anna Marie won the Queen of the Plough in 2008, at the National Ploughing Championships in Kilkenny. (Alf Harvey)

With my seven-month-old grandson, Seán Óg, when I received an honorary doctorate of philosophy from Dublin Institute of Technology in 2006.
(Lensmen)

My husband John with his beloved pipe.

Before we left home for Anna Marie's wedding to Declan Buttle, in June 2008.
(Gibbons Photography)

Anna Marie, Declan and our extended family at their wedding in 2008. (Gibbons Photography)

My sister Eileen's husband, Tommy Brennan (*left*), with my husband John in their favourite spot: sitting outside the NPA headquarters and watching the world go by, at the 1995 Ploughing in Ballacolla, County Laois.

John and me with our first grandchild, Seán Óg.

Ready to dance the night away with my sisters Eileen (*left*) and Betty, at a wedding in 2014.

With my dear friend Sister Consilio, founder of the Cuan Mhuire addiction centre, in 2016. Sister Consilio is also a very fine baker and she made my wedding cake in 1966. (Peadar Doogue)

DJ and his first son, Seán Óg, at a Laois County Ploughing Match in 2008. (Alf Harvey)

Putting on the Passion Play in Ballylinan has been one of the highlights of my life. Here I am with Sheila Graham (*middle*), who produced the play up to 2010, and my sister-in-law Mary Brennan, who played Mary, mother of Jesus, in the early years and also soldiered with me in the NPA for thirty-three years. (Alf Harvey)

The cast and some of the production crew after another sell-out performance in Ballylinan Hall in 2010. (Alf Harvey)

I attend many events for work, but there's nothing like a family celebration. Here I am at my grandson Tadhg's First Holy Communion in St Anne's Church, Ballylinan, in May 2017. Tadhg is to my right, his older brother Seán Óg is on my left and his sister Dearbhla and cousin Saran are in front.

Reliving my camogie days with my four grandchildren, Dearbhla, Saran, Tadhg and Seán Óg.

Anna Marie, her husband Declan and DJ and his wife Fiona helped me to celebrate when the NUI presented me with an honorary doctorate of law in 2014. (Lensmen)

Showing my youngest grandchildren, Dearbhla and Saran, the workings of the plough in Ratheniska, County Laois. (Alf Harvey)

I really enjoy being surrounded by family. Here we are celebrating fifty years of the Laois Ploughing Association, near Cullohill, County Laois, in February 2017. Former Queen of the Plough, my sister Eileen, is happy in the driver's seat. (Alf Harvey)

Where I am happiest, at home in Fallaghmore with Anna Marie and DJ.

Who can make 100,000 folk
Face out into the wild?
On three days in September
Into barren fields they filed
I know that Cromwell did it once
He had little else to do
But there's more crack in going ploughing
With Anna May McHugh

She can fill up five big tented bars
With farmers 'til they burst
And turn rows of lea and stubble right down into dust
She seems to be perfecting the record traffic queue
You're right, that's her and the NPA
Bold Anna May McHugh

Who can bring the Queen of Ireland
Mary R from Phoenix Park?
To small villages in Ireland
Lines of ministerial mercs?
Who can speak on television, and on radio miss no cue?
You're right, she can do everything
Bold Anna May McHugh

If she could run for Taoiseach
She would win it in a hack
If she ran the country like the plough,
Sure millions would come back
But she's far too good for politics
She knows more than all that crew
So she's sticking to the ploughing
Our Anna May McHugh

And when she goes to heaven
There'll be stubble in each cloud
And Esso and Kverneland will be embroidered on her shroud
She'll be up with Buddy Holly
Who sang of Peggy Sue
But me, I'll save my vocal cords
For Anna May McHugh

How could anyone not laugh at that? It's lovely to think that the Ploughing inspired so many songs. I think it highlights the warmth Irish people have for the event. I have such great time for the members of our Association and we've made lifelong friends with so many people. There isn't a corner of Ireland where I could get a puncture that I wouldn't know someone to call for help. And they would all be Ploughing people. There are people I could ring in the middle of the night and say: 'Can you do something for me at 8.00 am?' and they would say, 'No problem, Anna May. Just tell me what you want.'

9. A year in the life of the Ploughing

When people are sauntering around, enjoying the sights at the National Ploughing Championships, they have no idea that we are already planning the following year's Championships. Running an event of this size means you have to plan well in advance. The mantra at the back of my mind is always that one bad event could close us down, so it's like planning a military operation. Nothing can be left to chance. And if any activity or stand-holder causes problems, they are not entertained the following year.

My home in Fallaghmore is where it all happens. It looks like a typical farmhouse when you drive by, but it's the nerve centre for all things Ploughing. When I started working from home, my desk was the dining-room table. Then the Ploughing grew and you couldn't see the table for papers. We built an extension with a separate entrance in 1995. That houses two offices downstairs, and two upstairs. But the Ploughing continued to grow and we needed even more space. In 2013, we erected our wooden cabin. It has two offices and a little kitchen. And would you believe it, we might need to extend that soon. We now have five full-time staff all year round, as well as a team of five seasonal staff. You might wonder what we are all doing, given that it's just one event, once a year. To understand the logistics involved, it might help if I took you through a typical year in the life of the Ploughing. We held last year's National Ploughing Championships in Screggan, Tullamore, County Offaly, and we'll be back there again in 2017, so the 2016 event is a good one to look at.

We can make no preparations for an event until we have the site, and getting the perfect site for the Ploughing is not an easy task. I have been bundled into the back of jeeps and told to wear sunglasses when I've been looking at sites around the country. The US Secret Service wouldn't be as good as some of the ploughing people at organizing these visits. I suppose the potential hosts don't want the word getting out in case we don't go ahead with their site. They won't want their neighbours knowing that they were turned down for the Ploughing. And, to their credit, the site might be perfect but the road network might not suit.

People think the land is the most important issue, but in fact the road structure is probably the number one factor in choosing a site. You could have the finest farm in the country but if you don't have the network of roads close to a motorway, then it's going to be a nightmare for traffic. Delays are going to be inevitable with an event this big, but if you have to queue for an hour to turn into a field, you're not going to be very positively disposed towards the Ploughing by the time you get out of your car. And if you are sitting in your vehicle for hours trying to go home, you might not be in a big rush to return the following year.

We like to have the site pinned down for more than a year before the event. Although that does not always happen. There were a few sites in contention for last year's Ploughing, but in October 2015 we were still looking for somewhere for the 2016 Ploughing. We were considering sites in Kilkenny and Carlow. We were actually on one of the locations under consideration when we got a phone call from our previous chairman, P. J. Lynam, and our Offaly director, Michael Mahon, saying they were standing on a lovely site in Screggan, near Tullamore. That's generally how it happens. There isn't a formal application process for landowners who wish to offer their farms for the Ploughing. Usually, the county ploughing associations would approach us and say there's a site worth looking at in such a place. They'd have driven around it and checked it out before they would recommend it. Mind you, the site we got for the 2004 and the 2006 Ploughing in Tullow, County Carlow, was spotted when we were just driving around an area. Someone looked out at the land and said: 'Wouldn't that be a lovely site for the Ploughing?' Kevin and Margaret Nolan and the Byrne family owned that land, and indeed it was a lovely venue for the Ploughing. My oldest brother, JJ, was always a wonderful help when it came to sourcing sites for the Ploughing. He got very interested in ploughing after I took up this job and he would accompany me to look at farms. He had a great eye for picking a good site and was always very supportive in my work. He was the typical big brother – very good for advising me if I ever came to him with a problem.

As soon as we got the call about the Screggan site, we hightailed it to Tullamore and met the site owner, Joe Grogan. He was very enthusiastic and is a great supporter of the Ploughing, so I got the feeling that this could be the site for us. Another key consideration in choosing a site is ensuring that there are no overhead wires above the exhibitors' area. You can plough under overhead wires, but you can't have stands under them for safety reasons. There were absolutely no overhead wires on the host farm so that was good news.

Our next move was to get a guarantee that we could have the use of the Tullamore bypass for Ploughing traffic only. It took about a month to get the agreement from the National Roads Authority to close the road but, in the meantime, we were talking to neighbours to establish if they would provide fields for car parks. Normally you are only dealing with the host farmer and a few landowners, but in Tullamore we had to draw up contracts with more than twenty landowners because there were so many smaller fields around the area. In the end, we had secured a site of 800 acres. You would really need between 650 and 700 acres to comfortably host the event. You need about 180 acres for ploughing, 100 acres for trade stands and exhibition space, 400 acres for car parking and a demonstration area of approximately 25 acres. I know 400 acres sounds like a lot for car parking, but we must have a Plan B in the event of bad weather and car parks becoming unusable. We always secure additional parking further away from the main car parks.

People are constantly curious about how much money the landowners get for hosting the Ploughing. It won't make them millionaires, I can tell you. It is quite an honour to host the Ploughing but it won't make you up, financially. Of course, farmers have to be compensated because they lose the use of their land for several months. And they lose a bit of privacy too in the weeks before and during the Ploughing.

We've never differed with a host farmer over the compensation, and I honestly feel I could go back to any of the site owners we have dealt with over the years and work with them again. We pay depending on the acreage and give a higher rate for grassland compared with stubble ground – land that has been harvested for crops. It's mostly the neighbouring farmers who provide the fields for car parking. You might have a landowner who is looking for an outrageous price to use his field for additional car parking – you'd almost buy the land for the price asked – but I wouldn't give in to that. We are adamant that every farmer gets the same price for the land used for car parking, so in that case we'd just move on. I think most landowners respect the NPA because of our reputation for being straight, and please God that will always stand to us.

It cost exactly €5,144,510 to run the National Ploughing Championships in Tullamore in 2016. A bit more than the £9 3s. 5d. it cost to run that first event in Captain Hosie's field in Athy, back in 1931. The biggest expense of a Ploughing Championship is not the land. Not even close. It's the steel trackway that we use to create roadways so that machinery, cars and people can get around the site with ease and without sinking into the ground if it's wet. The trackway costs up to €700,000 per year, but without it we'd be lost. It kills me to see that money going out of the country every year because we can't get anyone in Ireland to supply the trackway. We first started using it in the early 1980s, and that was thanks to an exhibitor who had been to a show in England and saw it there. Before that, we were constantly getting phone calls saying a tractor was stuck to the axle in the gateway, or the car parks were in great difficulty.

We got a very small amount of trackway from England to try it out, and we were amazed at how effective it was. You'd never think to look at it, but it can withstand pressure of 25 or 30 tonnes. No matter how bad the weather is, it will hold up. We tried to get someone to supply it in Ireland, many years ago, but they said the demand just wasn't there and it wouldn't be worth their while. The trackways are more popular in England, France and Germany, for all sorts of outdoor shows and events, so we have no problem sourcing it overseas. Every year we have been getting more and more trackway. Some 25 kilometres of trackway was laid for the 2016 event, and in 2017 we'll have 30 kilometres covered, just to make it easier for people to move through the site.

The next biggest expense is the electricity and getting the electrical cabling in place underground. We had 65 kilometres of electrical cabling last year and eleven generators supplying electricity. I remember when one generator was more than enough. It takes a team of between thirty-five and forty electricians to prepare the site for us, and they really are working flat-out for the few days before the Championships begin. Water supply is another big cost. The payment to landowners comes far down the list.

The Screggan site ticked all the boxes for access. Another bonus was that visitors could take the train to Tullamore and get a Bus Éireann shuttle bus for the remaining 5-kilometre trip to the site. Once we had secured the site and got the road access sorted, we held a public meeting in April with the neighbours to explain what was involved. It's very important to take the neighbours' concerns into account when planning the event. They still have to go about their business, so we have residents' passes to make sure they are not stuck in traffic getting to work. Some of the residents in Mucklagh, beside the site, were really nervous about the Ploughing, but we were able to reassure them, with the help of the Gardaí. We also gave them the names of the host farmer and the neighbouring farmers for the 2015 event so they could ask them any questions they might have. And we provided security at the entrance to their housing estates during the days of the Ploughing to ease their minds. Afterwards, we had a review meeting and a few of them said: 'God, we were so nervous about the Ploughing coming here and yet we went every day to it. We couldn't keep away from it!'

In Tullamore we had people trying to get to work in the hospital, trying to get their children to the crèche, lorries trying to deliver gravel and the like. Funny enough, I remember talking to a nurse after the event and she said she never got into work as quick as she did during the Ploughing because the whole traffic plan was in place. Gardaí also went to the local schools and suggested they might close as it was an educational event. The schools agreed with this, which also lessened the morning traffic. I know it doesn't suit everyone to have the Ploughing on their doorstep, and we recognize that by giving complimentary tickets to the immediate neighbours, as a goodwill gesture.

The admission prices have stayed put for a very long time and we have no intention of increasing that. Adults are €20. Students and OAPs (people over 66 years) are €15, and there are discounts for pre-booking. Children under 12 are free. We feel it is very important to keep the outing affordable for the general public.

The information packs for exhibitors went out in March 2016 and from then on the pace really picked up. Exhibitor space is sold by the metre. This year, the early-bird bookers get a 9-metre deep outdoor stand for €156 per metre frontage, while the 18-metre deep stands are €176 per metre frontage. Indoor exhibition space costs €1,100 per unit for the early bookers. We had to increase the charges by 10 per cent last year, for the first time since 2006, because our own costs have soared in recent years. The cost of preparing the site and providing the necessary facilities has really escalated and we could no longer absorb it. But it didn't deter exhibitors. It seemed like everyone was booking early last year as more than 90 per cent of our stand space was booked before the end of May. Bookings were almost 30 per cent up on the same time in 2015, which was unprecedented for us. I suppose it was down to a combination of factors: the economy improving, the growing interest in the event and the increasing clamour for the best location. Exhibitors were keen to get in early so that they had the best possible spot.

I have to acknowledge the support that the exhibitors give us, year after year. I know they don't come for the good of their health – they are running businesses and it makes sense to meet their customers at the event – but it's not always easy for them. The weather might be bad, or the venue mightn't suit and yet they return year after year. We always give them a good listening ear when it comes to feedback because we wouldn't have the event without them. I am very conscious now of not letting the Ploughing get any bigger, and the NPA council would feel the same. Besides, it would be very difficult to get sites big enough, if it kept expanding. And if it gets too big, we mightn't be able to look after the exhibitors that we already have. Some of them are with us for up to forty years and they deserve to get our attention. We had almost 1,700 stands last year and I don't know if we could look after one more stand. You can imagine that many people trying to get into a big field in September and the logistical nightmares involved. I remember climbing two storeys high on a stand last year and looking across the site at the thousands of people milling around. I said to myself: 'Lord, it's frightening. Anything could happen and I'm responsible for it all.' In 2016 some exhibitors came from as far away as Australia, New Zealand and Canada. In fact, we had delegations from fifteen countries in Screggan.

We moved onto the site in early August and our architect, John Burgess, started the major task of laying out the site. I can recall doing this work when we had only twenty or twenty-five stands, back in the 1950s. It didn't take too long. If you were to tell me that we'd be laying out 1,700 stands, sixty-five years later, I'd have laughed at you. People often ask how did we manage to increase the size of the exhibition so much. It might be something to do with the fact that I always watched the advertisements in the farming press, like the *Farmers' Journal* and *Farming Independent*, and if I came across companies that were advertising agricultural machinery, I would always send them literature on the Ploughing Championships. Nine times out of ten they would come back and say they'd like to exhibit at the Ploughing. We were determined to keep marketing ourselves, reminding people that we were there. After a while, they came looking for us, rather than the other way around.

As soon as John Burgess's plan was finalized, the trackway went down and it was full swing ahead for the next month-and-a-half. What were we doing, you might well ask? Apart from the trackway, electrical cables had to be buried underground, exhibition space had to be constructed and the water supply had to be sorted. And all the while we were liaising with the local emergency services, county authorities, safety officers, fire officers, HSE, road engineers, council officials, An Garda and environmental health officers. It is really like building a town from scratch.

Setting up medical services is another key part of our site preparation. We had seven first aid centres on the site in 2016, as well as a huge medical centre, and you'd need them all, with that number of people coming to the Ploughing. Despite our emphasis on safety, we cannot prevent people from having trips and falls or feeling unwell. Sadly, we've had two deaths at the Ploughing. A ploughing competitor, Willie Gibson from Cloughjordan, County Tipperary, had a heart attack as he was competing at the Ploughing in Carlow in 1990 and he died suddenly. The ambulance was there in seconds, but there was nothing they could do.

Another man died as he was walking on the trackway on the first day of the Ploughing in Athy one year. I think it was a heart attack. When something happens at the Ploughing, our first thought is always: 'Did we do enough? Was there something we could have done to avoid it? Is there something we could do to minimize the risk in the future?' We want to leave nothing to chance.

We were very saddened to hear that a farmer died on the train, on his way home from the Ploughing, on the opening day of the Ploughing in 2016. But I was glad to read in the papers that the Iarnród Éireann staff and passengers had done everything they could to help him and people had prayed aloud for him. It must have been some comfort to his relatives and friends to hear that.

But the Ploughing also saved at least one life, in Ratheniska, County Laois, in 2015. A man had a heart attack and collapsed, but luckily there was a member of security nearby and he brought him back to life. He was then swiftly assisted by the Order of Malta. They said that if it had happened at home, he might not have survived because he wouldn't have received treatment in time. We always have a few cardiac cases at the Ploughing and we see a lot of dehydration. Another year a very young baby became quite ill after arriving at the Ploughing. The baby was suffering from asthma and, by pure fluke, some Order of Malta people were walking through the car park and they resuscitated the child. We were very relieved to hear that.

On a positive note, we've also had marriages, and presumably births, as a result of the Ploughing as it has brought many people together. Just recently I met a man at a funeral who told me that he had met his wife at the 2012 ploughing. I am frequently being told of couples who met at the Ploughing.

For example, the soccer player Kevin Doyle met his wife, Jennifer Harney, at the Ploughing when they were teenagers.

The Ploughing in Athy was also the backdrop to an extra-marital affair, I'm sorry to say, but thankfully it was just a fictional one. *Fair City* character Jo (Rachel Sarah Murphy) was having an affair with Tommy (Geoff Minogue) and they decided to go to the Ploughing together in Athy in 2011, thinking that their spouses would never find out. But as everyone knows, you'll always be seen by someone at the Ploughing. Sure enough, the lovebirds were spotted and the rural outing came back to haunt them! Many's the time the two characters cursed the Ploughing afterwards. I don't normally like people cursing the Ploughing, but I let them get away with it.

And an even more famous actor also has a link with the Ploughing. I remember hearing how the *Farmers' Journal* news editor Patrick Donohoe buttonholed the actor Russell Crowe at a film premiere in Dublin in 2014 because he had heard about the actor's farm in New South Wales, Australia. The *Gladiator* star is a farmer's son and keeps a herd of Angus cattle on his ranch. As soon as he heard Patrick was from the *Farmers' Journal*, the actor was delighted to be able to tell him that the Ploughing Championships were coming up in September. I'm sure he'd enjoy a day out at the Ploughing, and we'd facilitate him in any way he liked.

But of course it's not all sweetness and light at the Ploughing. As you can imagine, running an event this size has brought more than a few challenges and dilemmas our way. Over the years there have been efforts to hold various protests at the Ploughing over certain issues. I suppose people think they have a ready-made crowd and lots of media to report on their grievance. But apart from one major incident, which I'll tell you about later, we've never entertained that. I remember one man with a placard protesting outside a bank's stand at the Ploughing, over the way he had been treated, and I had to go down and talk to him. I said to him: 'Would you mind leaving it to another day because this is the Ploughing?' and he said: 'Anna May, are you asking me to leave or are you telling me?' I said I was asking him, and he said that was no problem, but if I was telling him, he wouldn't go. And he left quietly with no fuss.

You would get rumours about farm organizations planning to protest over something the Minister for Agriculture has done. I would have to ring them to get an assurance that there would be no demonstration. And I don't mind saying that I would tell them that they would never enjoy the same central location at the Ploughing if they did a thing like that. But we share a great mutual respect with the farm organizations and my family have been members of the IFA for years. We all share the same objectives. But we have to be firm with everyone because we don't want the Ploughing hijacked by groups for their own purposes. Otherwise, where would it end? You can't lose control of an event this big or anything could happen.

I suppose we learned our lesson many years ago when we were placed in a very awkward situation involving a protest. We didn't put the foot down and forbid the protest, and I never want to face such a scenario again.

I had to withdraw an invitation to the Minister of State for Agriculture, Paddy Lalor, to visit the 1966 Ploughing in Wellington Bridge, County Wexford. (Back then the post was known as parliamentary secretary.) We had been told that the National Farmers' Association, now the IFA, was going to have a big march if a Government representative turned up. The Ploughing was being held during a very controversial time for farmer/Government relations because, just two weeks earlier, Minister for Agriculture Charlie Haughey had refused to meet the NFA deputation when they'd marched from Bantry to Dublin in the Farmers' Rights Campaign. Farmers had endured a very tough time with the collapse of livestock prices and milk prices and they were looking for the right to negotiate with the Government as a representative group. Not only did he refuse to meet them, Mr Haughey had added fuel to the fire by referring to them as 'pipsqueaks'. His stance led to farmers sitting on the steps of Government Buildings in protest. The sit-in would eventually end, after twenty-one days, when Mr Haughey agreed to meet them, along with the taoiseach, Seán Lemass. (Although the dispute would continue on for months and would result in the jailing of several farmers, including one of the NPA's former chairmen, Michael Mahon from Offaly)

But back to the sit-in. Unfortunately for us, the Ploughing was taking place when this was going on, which made things very awkward. Paddy Lalor had been invited before this stand-off arose and, despite our entreaties, the NFA was adamant it would hold the protest if a Government representative arrived in Wellington Bridge. It wouldn't have been a nice occasion for him, or for the NPA, so we felt we were left with no option but to withdraw the invitation.

It fell to me to go to the late Paddy Lalor and tell him we were withdrawing the invitation. Now I knew him, which probably made it worse, because he was from Abbeyleix. Paddy played hurling for Laois and he went on to have a long political career as a TD, a minister, chief whip, and member of the European Parliament. He was a very honest politician, and highly, highly respected.

I'll never forget having to go to his office to give him the news. It was a terrible thing to have to do and we've never done it since, nor will we. I won't tell you what Paddy Lalor said because we couldn't print it. He was very annoyed, to say the least. And it's no wonder. Telling a junior minister he couldn't come to an event was a shocking thing. And you can imagine Charlie Haughey's reaction when Paddy told him what had happened. I'd say the language was choice. I must say, I always found Mr Haughey to be a very good, co-operative minister and a highly intelligent man.

From that day on, I vowed that we would never be put in a similar situation again. Everyone has their issues, but they have fifty-one weeks of the year to ventilate them. There's no need to pick the week of the Ploughing, when we are just trying to run the event and keep our patrons, competitors and exhibitors happy.

Now, back to Tullamore again. On the afternoon before the opening day, the site got so busy that Chief Superintendent John Scanlon had to direct traffic and close off access for any additional cars. The site was as busy as most other shows would be on their 'show day'.

I had to leave the site at about 7.30 pm and master a quick change of clothes in the car while being driven to the Enterprise Ireland pre-Ploughing dinner, which was hosting international buyers from fifteen countries. As I left, I had no doubt that our site manager Tom Kelly and roads manager Billy Gray would continue working over and above the call of duty to sort out any potential problems before the morning. Everyone just puts their shoulders to the wheel at this stage to ensure that we are ready to face the crowds. Some of the team even work through the night, if necessary.

So, by the time the opening day of the 2016 Championships came along, on Tuesday, 20 September, we were ready for every eventuality and raring to go. I've honestly never regretted taking this job, but every year, between 8.00 pm on Monday night and 6.00 am on Tuesday morning – the opening day of the Ploughing – I always say to myself: 'Life could have been a lot easier than this.' Everything is done yet anything can happen with the weather or the traffic, or accidents, and I hardly sleep at all that night. I'd be listening out for rain or wind. I know it's out of my hands, but I can't help it. I'm known for pulling back the curtains at 4.00 am on the morning of the Ploughing to see the weather. And I'd be so happy if the weather was dry. I always send money to the Poor Clares in Graiguecullen the week before the Ploughing for good weather and to keep people safe. Without fail this little bundle of scapulars, medals and prayers comes back from the nuns and they give me great comfort.

The prayers worked for Screggan last year and the weather held up for the opening day of the Championships. Ploughing day starts for many of our team shortly after 4.00 am, particularly for the gate teams, all of which are led by volunteers. They head for the site well before first light to get everything in place for the early arrivals. Gardaí and the NPA control team are communicating on their radios from 5.15 am and you really would not believe how alive the site is by 6.00 am. Car parks are starting to fill, livestock owners are feeding animals, caterers are getting breakfasts ready and there is no going back at that stage. The event just takes off and our event management plan takes centre-stage, with particular emphasis on health and safety. Meetings are held every day with the emergency services to review how things are going and our control room is in contact with all areas of the event. We have more than 200 radio phones across the site and our site crews are constantly on the move, carrying out site repairs where required.

The whole enterprise is a huge logistical feat and a massive team effort. Every link in the chain has an important part to play. To be honest, that means my main role is to meet and greet visitors. From time to time I will be called on to make a final decision on something, but I have to trust the team that I have built around me. The only way they will feel a sense of ownership about the event is if I let them do their job without me looking over their shoulders. The office staff, led by Louise Brennan, Mary Claire Brennan, Geraldine Hooban, Ann Meaney, Anna Brennan, Ann Siney, Maria Mulhall and Carmel Brennan, open the NPA headquarters on-site by 6.00 am, and I can honestly say nobody will sit down again until they get dinner that night, probably at 10.00 pm, if they are lucky.

Some 100,000 people turned out that first day in Screggan – a record first day for us. I remember Joe Grogan, the site owner, telling me that he was watching the cars filing in that morning and he felt such a sense of responsibility that everything would go well. He was a marvellous site owner and one of the most co-operative we've ever had. I could have asked him at any point to get me a tractor or an excavator and he'd be back before you knew it with the machine, asking: 'What do you want next, Anna May?'

We had about 800 people working for the NPA at Screggan, between staff, stewards, judges and all the other volunteers. Another 800 people were working on the catering side. We heard afterwards that about 80,000 cups of tea and coffee were drunk during the three days of the Ploughing and almost 40,000 breakfasts consumed. Ploughing is hungry work, you know. When I started with the Ploughing in the early 1950s, you'd count yourself lucky to get a curled-up ham sandwich. Now you can get a full, sit-down, four-course meal with wine, food from around the globe and every type of coffee concoction you could dream of.

As the day comes to a close, the site crew usually meets at around 9.00 pm for a bite to eat in the site office and to report if all areas of the site have been cleared of visitors. It has become an annual event now that there will be one person, usually someone aged between sixty and seventy years, who has not arrived home or met up with their lift. In league with the gardaí, we trigger an emergency plan to try to locate the person. Thankfully, that has always ended up in a good news story, with the person invariably found on the train, or in the local pub at home, or gone for tea somewhere. The site crew then gets back to its plan for night repairs, so that the site is ready for the arrival of the public the following morning. You could hear of a person being phoned at 2.00 am to come to the site to rectify a burst pipe, or a whole section of track way may have to be moved because of ground conditions.

Local knowledge is always an integral part of the Ploughing. I remember a few years ago we had such a deluge of rain one night that you could honestly swim in a pond that was created in a dip in the ground at the livestock arena. We had to call a local farmer with a slurry tanker and he came in and sucked out the pond. We filled the dip with bark mulch and everything was in order for the next day.

You probably wouldn't expect that one of our annual jobs is helping to reunite people with their cars. We have huge signs at every car park with various animals on them to help people remember where they have parked, but some folk still forget which way they came in. Many's the night we've been driving around car parks, pressing the car key fob to see if the car lights up somewhere. I remember one journalist gave up trying to find his car and got a lift home with someone else once. We were coming in the next morning at 6.00 am and spotted this one lone car over in the corner of a huge field. So we were able to tell him where it was when he arrived.

The press room is another area that has increased massively since the early days. Back when I started working with the Ploughing, you could count the journalists attending the Championships on one hand, and you'd still have a few fingers left over. Now, would you believe it, we send out nearly 800 press passes every year. A newspaper or radio station might send a different reporter every day and each person needs their own press pass. We have journalists working in print, radio, video, television and online news and we also attract a few social media bloggers, too. My God, the scrum for sockets and desk space gets worse every year. I had a funny experience a while ago when a camera crew of three young people asked to follow me for a few hours at the Ploughing. I said that they were welcome to, but I had a fairly hefty schedule. That didn't faze them so we headed off. About forty-five minutes later they all arrived back at the NPA HQ asking did anyone see me. I was lost in action already, and I was over eighty at that stage.

Another time the rain was really quite bad and I was about to do a TV interview when the interviewer said I would have to use an umbrella, which does not look great, but what can we do? I said nothing but took my place and when the camera went live, I just dropped the umbrella and talked away like the sun was beaming. I may have been soaked, but the picture looked good.

As well as the Irish media, we have a few journalists coming from overseas every year. They mostly come from agricultural publications in the UK, France, Germany, Austria and a few other European countries. One year we hosted some Chinese journalists who had come with a trade delegation. They were so impressed by the fact that it was an outdoor event because they were only used to big indoor exhibitions. And we had another first in Screggan when the tractor football appeared on Fox News in the USA. The Ploughing reaches far and wide!

There is no sweeter feeling than leaving the Ploughing Championships on the last evening when everything has gone well. It's like two concrete blocks have been lifted off your shoulders. That's exactly how I felt after Tullamore. The site still has to be cleaned up, but you're just dealing with exhibitors and everyone is much more relaxed.

I'm always amazed at how often the Ploughing is referred to, in passing, on the radio or television during the year. You never know when it might pop up. The presenter could be talking about running some big or complicated event, like the Web Summit or an awards ceremony, and they'd say: 'The organizers could learn a thing or two from the NPA on how to run a major event.' I get a great kick out of that and feel like I grow a good inch taller when I hear the Ploughing referenced like that!

I like to keep in touch with how other outdoor events are organized because I think we can always learn something. I remember going to the Oxegen music festival at Punchestown Racecourse in 2010, a few days after Anna Marie's son, Saran, was born. I headed straight to the hospital to see them after the festival, so I was still wearing my muddy boots. I got a few funny looks because of the boots and when I told people where I'd been, the looks got even funnier. They were probably wondering what an old one like me was doing at a music festival. But I think it's always interesting to look at other outdoor events to see what works, and what definitely doesn't work. What I learned from Oxegen was that the girls were wearing more on their feet than on their bodies and there was all sorts of music thumping but I couldn't name one of the bands. All I remember is lots of young people enjoying themselves, and being very impressed by their first-aid set-up and their communications centre.

I've been to Electric Picnic in Stradbally a few times too and found it to be a friendly, relaxed event and very family-oriented. The last field of the Electric Picnic car park was next to our site for the Ploughing at Ratheniska, so Stradbally is well used to dealing with people in wellingtons.

The 2016 Championships proved to be our most successful, attracting 283,000 visitors over the three days. We didn't have to think too hard about returning to the site this year. I love the days after the Ploughing when you get to hear all the stories about the things that happened: the almost disasters that were averted; the ploughman who believed in his lucky ploughing boots; the steward who brought a camera in his pocket to make sure he got a picture of an offending competitor breaking the rules; the judge who never turned up to judge, but three hours later remembered when he saw the judging team finishing the task; the parents who had to drive two or three hours to collect a teenager who missed the bus home.

We eventually left the Screggan site in the early days of November. Yes, it takes more than one month to return the site to its original condition. In the past I spent many long days picking up litter with a team of helpers, long after the event was over. You wouldn't believe the amount of bags we'd fill. And that's on a site that's covered with bins. Some people are so bad that if you don't leave the lid of the bin open, they'll just drop it on the ground beside the bin. Litter has to be collected quickly because if there is wind, it carries the rubbish everywhere.

I don't pick up litter now because we have a company looking after that for us. But I didn't mind it so much. I found it a great way of clearing my head after the event. If we were on the roadside, though, I'd have to disguise myself in case anyone saw me. There was a lot of roadside at the site in Edenderry in 1982 and our first job after the Ploughing was to clear the litter from the ditches, because you didn't want passers-by to see litter and blame the Ploughing for ruining the area. I used to laugh to myself, thinking of the people who would have said I had a very glamorous time of it, up on the bandstand, mixing with the VIPs. Some people definitely think I live the high life. We were up at the Northern Ireland Ploughing Championships a few years ago and Anna Marie got talking to a standholder as she was walking around. The conversation came around to our own Ploughing event, but she didn't say who she was. He went on to say 'that McHugh woman' had flown in to the Championships that morning. 'She came by helicopter with her daughter and grandson this morning. Sure, those people don't travel by car at all,' he told Anna Marie. She was really enjoying this and she delighted in telling him that she drove us to the event and she'd never been in a helicopter in her life. He went puce in the face.

I was always very hands-on with the Ploughing because I believe you shouldn't ask people to do something that you wouldn't do yourself. In 1991 I found myself on the site of the Ploughing in Crecora, County Limerick, at 7.00 am, rolling around a bale of straw in front of the NPA marquee with Maurice McEnery, our then site manager. The ground had been very wet and we were laying down the straw to dry it up. Nothing glamorous about wellingtons and muck, I can tell you!

Screggan was our 85th National Ploughing Championships. I've included all the locations for the Championships in an appendix at the back of this book and if you look at the list, you'll notice how the Ploughing was very adventurous in the early years, taking in venues as far flung as Wicklow, Roscommon, Tipperary, Wexford and Kerry. We could do that when the event was small because a good road network wasn't crucial. Now we have no choice; we have to favour central locations like Tullamore, Carlow, Laois, Kilkenny and Kildare.

Donegal would love to host the Ploughing and the ploughing organization is so professional up there. Without doubt, they would do a terrific job. But unfortunately the distance would put a lot of people off. Can you imagine bringing exhibitors from Kerry to Donegal? The distance alone would deter them from going.

Cork and Kerry people are wonderful supporters of the Ploughing. I think all of Cork and half of Kerry come to the Ploughing and we have had the Ploughing there in the past. But, like Donegal, if we went down there for the Ploughing now, we would be asking a lot of people to travel very long distances. Our books clearly show that when we have a central location, people are happier and more money is spent.

We were in the west a few times in the early days. Boyle, County Roscommon, was a washout in 1957 and we've been to Athenry in Galway a few times, but I couldn't see the Ploughing going west of the Shannon again. We looked at sites in Longford in recent years, but the fields were not ideal for us and the road network wasn't suitable.

We're often asked why we wouldn't buy a permanent site. I'm not against the idea at all. In fact, it would be so much easier for us because you wouldn't be starting afresh every year. You could put up permanent buildings. You wouldn't have to spend the enormous amount of money we spend on trackways. You wouldn't have to do all that work putting electrical cables in the ground every year, getting water supply, bringing toilets. All those things could be built into a permanent site, and that would be a huge advantage.

And yet the novelty of going around from county to county is a wonderful attraction. People love to come to a different place to see how farming is done and to stay in different hotels. You're spreading the business around the country, too. Exhibitors also say they sell more if the Championships are not in their own county.

Having the Ploughing in an area can be an inconvenience to people in the locality, but at least they know it's just for a year or two and it will move on again. If you have a permanent site, there's a danger that you could become a permanent inconvenience. You would also have to find a site that has enough land to rotate for competitions. And then you'd have to farm the land, or lease it for the year. You'd have to be very careful to lease it to someone who would farm it the way we'd want, so that it would be in tip-top shape for the Ploughing. There are so many pros and cons.

I wouldn't rule out a permanent site, if we got a place where neighbours would row in behind us. If a massive place came on the market in a central location, perhaps we'd have to think about it. I don't know what the other directors think, though. It's something that would have to get the backing of everyone. Maybe someone with a love of plough-ing will leave us a site in their will! That would help to concentrate the mind.

So that's a quick behind-the-scenes look at some of the work that goes into staging the National Ploughing Championships. Maybe you can now see why I live and breathe the Ploughing every day. But I wouldn't want it any other way. After sixty-six years working with the Association, there's probably as much brown soil running in my veins as blood.

10. A woman in a man's world

People often say it must have been difficult running the National Ploughing Association while rearing a family. But I'll tell you something now that not many working women could say. I never had to take a week off because of a sick child or a family emergency. That's the beauty of having your office in your home. If I didn't get the work done during the day, I got it done at night when DJ and Anna Marie were in bed. I worked late into the night a lot, and I still do. It's amazing what you can do when you have to do it. Of course, I was very lucky in that regard. If I had a job where I had to clock into an office in Dublin every morning, it would have been a different story entirely.

And you have to remember that the Ploughing was not as demanding when DJ and Anna Marie were born. They grew with it, or it grew with them, I suppose you could say. If I was starting again, it would be a totally different story. The Ploughing is such a huge operation now that you couldn't possibly be at the helm without full-time childcare. The children were always number one and the Ploughing came next. Having your own family nearby is a wonderful back-up for any working parents. My sisters Betty and Eileen were a great help with the children when they were young, particularly Eileen, who was always around.

But John was the linchpin holding it all together. I couldn't have done it without his support. And isn't it the same for men who have demanding jobs? They couldn't do it without someone keeping the show on the road at home. It's a well-known fact that behind every successful woman is a good man. If I knew that John was contrary or annoyed about me going to a meeting at night, it would have made things much harder. Instead, he was with me all the way along, right until the day he passed away.

John was very, very good with the children. He took them everywhere with him and they loved him for it. If a child didn't take to John, they'd take to no one. He was great with children. He had huge big hands and I can still see them going around DJ when he was a tiny baby. It gave me a lovely feeling. I remember the carrycot on the back seat of the car and not even a seat belt on it. But that was the way things were at that time. There was no such thing as a car seat for children, or even seat belts in the back seats of cars. And if you heard about a booster, it was a booster vaccination for cattle, not a booster seat. John was a rock for me because he was always there whenever I had to go away. But he had all his chores on the farm, too, so I'd always have a girl in the house helping out.

For all those reasons, I have very little guilt about working when the children were young. I knew they were being well-cared for. But there was one occasion when work clashed with family and I still feel bad about it, to this day. I missed the Confirmation ceremony. Not one child's Confirmation, but two. I had to go to New Zealand for a month and a day for the World Ploughing in 1980. I had a feeling that it would clash with DJ's Confirmation as he was in sixth class. And indeed, it did. The Confirmation day fell right in the middle of the New Zealand trip. That was bad enough, but then the teacher told me that, for the first time, fifth and sixth class would be making their Confirmation together.

Anna Marie was in fifth class. I couldn't believe it. But what could I do? I couldn't cancel the trip.

I did everything I could before I left, laying out all their clothes, and John's clothes. My sisters Betty and Eileen and my sister-in-law Mary Brennan gave them a hand to get prepared on the morning. In fact, Mary made the lovely brown velvet outfit that Anna Marie wore for her Confirmation.

On the day of the Confirmation I might have been thousands of miles away in Christchurch, New Zealand, but my mind was definitely at home. One of the ploughing competitors, John Tracey from Carlow, also had two sons – Eamonn and Michael – making their Confirmation when we were gone, so we were in the same boat. I'm happy to report that John took silver at those Championships, so the abandonment of our children wasn't in vain. Nevertheless, Anna Marie still gently reminds me about it, to this day. And DJ claims that he needs counselling to get over it!

(Incidentally, John Tracey's trophy was a beautiful silver rose bowl that had been engraved with the names of the previous winners. It was such a big, round bowl that when he got home, he sat his little baby Sharon into it and took a picture. She couldn't have been more than a year old at the time. John's wife, Lil, showed me the photograph afterwards and we had a great laugh about it.)

By all accounts, DJ and Anna Marie had a great day of it and when I got back, we all got dressed up again and took a family photograph. I'm sure I brought them home something good to make up for being away. The children always did well out of those trips because I'd pick up things that we'd never seen in Ireland, like instamatic cameras and digital watches.

Two Confirmations in one house is a lot, but when I was a

child three of us made our Confirmation together – Paddy, Gerald and me. It was a big thing to get three children rigged out for that. I remember we were brought into Hadden's in Carlow for the clothes. The boys got suits and I got a white frock and a pink coat. I remember we got a mineral in the shop after the ceremony. That was a huge event for us.

That was the only family dilemma that truly stands out in my mind over all the years working with the NPA. Of course, we had a few minor mishaps along the way. My husband John was always charged with dressing the children for the opening day of the Ploughing as I would have already left for the venue. One year he made the fatal mistake of dressing Anna Marie in the lovely tweed outfit I'd left out for her, before he dressed himself. He was shaving and idly looking out the back door when a flash of something caught his eye. Anna Marie was only three years old at the time and she had taken it upon herself to go exploring. One thing led to another and she arrived back to the house covered in cow manure. The lovely frock was a write-off. John cobbled together another outfit and he never made the mistake of dressing her too early again.

I also remember myself and John bringing the children to the Spring Show one year, just as my parents had taken us so many times. Of course, I had the children dressed to kill, as any parent would when heading off for a big day out. We were sitting in traffic on the Grand Canal in Dublin and I heard a cry in the back of the car. When I looked around, wasn't DJ's immaculate white shirt spattered with blood? Anna Marie had fallen out with him and boxed him on the nose! We had to find the nearest drapery store and buy him another shirt before we could continue on our way. They loved the Spring Show, just as we did, and they always made a beeline for the Eason book and toy shop just inside the main entrance of the RDS. They are true children of the Ploughing, those two.

To be honest, I don't think being a woman did me any harm in my work. In fact, people probably remembered me better because I was a woman and there weren't too many of us in agricultural circles at that time. For many years, I've been the only woman at so many meetings and get-togethers. I can't stress how courteous the men were to me, because I was a woman. If I was a man, they probably wouldn't have been as nice. They always treated me with great respect, even down to carrying my cases if we were going somewhere. I never once felt excluded because I was a woman. I suppose knowing the people so well made it easier. And I never had trouble making conversation with people because you never run out of chat when you have ploughing in common. It would be a different story if you were walking into a function in Dublin where you knew no one and were struggling to find common ground with them.

But that has all changed in recent years and today our NPA office is totally staffed by women. Basically, there'd be no ploughing without the women. When I was younger, it was not uncommon for a man calling to a house to ask for 'the boss man' if a woman answered the door, even if he knew that he was talking to the woman of the house. It was just an accepted term. Similarly, men would ask for 'the boss man' when they came across women in day-to-day business or if a woman answered the phone. If a man was crazy enough to ring up and ask to speak to 'the boss man' today, he would get short shrift.

I think it's great to see women in such high-profile jobs now. I'm thinking of people like Mairead McGuinness, vice president of the European Parliament, who is doing so well. And there's Siobhán Talbot, incredibly capable as she runs the massive organization that is Glanbia. Fiona Muldoon is head of FBD Insurance and Bernie Brennan is president of the Royal Dublin Society. And Bord Bia now has its first female chief executive – Tara McCarthy. All strong, capable women and none of them appointed because they were the token woman.

Women are better at promoting themselves now than we were in my day. They definitely have to work harder than men to prove themselves but when they get their place, they can really prove their worth. People have told me I'm a trailblazer for women, but I never prided myself on doing that. I'm no better or worse than anyone else. I just went along and did my job to the best of my ability. And I enjoyed every minute of it.

I've often heard it said that I must be very tough to have risen to the top of the NPA and stayed there. The truth is that I'm not a bit tough, behind it all. Still, I suppose it does no harm if people think I am a bit of a Maggie Thatcher. I smile when I hear people saying: 'Oh Anna May's a tough person', or 'Don't ever cross Anna May or you'll be sorry.'

But you do have to be strong-minded in this job and hold your ground. For example, I'm very firm on alcohol at meetings. It's strictly banned. I remember one man coming into a meeting with a pint of beer and I said: 'Chairman, we are not allowing that.' And from then on, no one has ever brought drink into a meeting. And everyone will agree that our meetings are better for it, well-conducted and efficient. And the bar is there afterwards if anyone wants to have a drink.

I hate anyone trying to pull the wool over my eyes. If they get away with it once, they won't do it twice. We deal with a lot of contractors and, perhaps when they didn't know me, they might have tried to do that. They might have tried to charge a higher price because I was a woman, but I don't think they'd try it now.

I would always look for the best value and try to get a bit of a discount, so maybe I got the name of being a bit of an iron lady. But there must be give and take on both sides or you'll never strike a deal, and I do get on extremely well with the contractors.

Loyalty is very important to me. If people are loyal to me, I will be loyal to them, to the absolute limit. I am like a tiger protecting her cubs when it comes to the Ploughing. I would be very sensitive if someone tried to run down the event. It's like someone criticizing your child. You really would rather not hear it. The Ploughing is very close to my heart. I couldn't overemphasize how much it cuts me to the core when an article appears in a newspaper criticizing the Ploughing. It really kills me, especially if it is framed in a way that suggests that the NPA is all about grabbing cash or feathering our own nests.

I would say to a person who criticizes it: 'Look, that's your view, but it's not mine.' Sure, I would listen to any criticism and I believe people are entitled to their opinion, but I would defend the Ploughing to the last. I remember one man telling me he'd never exhibit at the Ploughing because it was nothing more than a car-boot sale. I thought that was a terrible insult. But I held myself back and I was very polite. I said I respected his opinion but I didn't share it, and thankfully not many other exhibitors or members of the public did either, and let's leave it so. I think he left me feeling sorry that he'd said such a thing, and no harm!

Part of the reason I'm so defensive about the Ploughing is

because I know the work that people put into it. Be they volunteer or paid staff, they all know that if you enlist, you must soldier, no matter what the conditions, and unfortunately conditions at the Ploughing can often be quite challenging. We really are one big family and while we might not see each other from one year to the next, we all just move to the ploughing site and everyone takes up their job where they left off last year and it goes on from there. Then we have people who join the team through our local organizing committee when the Ploughing comes to their area. They get sucked into the excitement and find themselves travelling with the event to the next venue and becoming part of the Ploughing roadshow. Lifelong friendships have grown from stewards working together during the Ploughing, or locals providing B&B. I have such great people around me and we couldn't do it without them. They would do anything for me, and I hope they know I would do the same for them. If you can help someone as you go along, do it and your living will not be in vain.

Carrie Acheson, who does all the public announcements at the Ploughing, often says there is one great thing about being part of an organization like the Ploughing – you are guaranteed a great crowd walking after your coffin. It always gives me a chuckle, but there's truth in it. We have become a real family through the Ploughing and we do give our own a great send-off.

While there's a great team behind me, I suppose it's only natural that the focus falls on me as the face of the organization. I feel I'm totally accepted as a female managing director now. Gender doesn't come into it at all, and nor should it. But of course I come under the same scrutiny as all other managing directors, be they male or female. Where once the media was excited about a woman leading a male-dominated organization, now they are more excited about how much money that woman is earning. Every so often the matter of the NPA's accounts comes up and reporters ask lots of questions about my salary. In fact, I'm told that if you put my name into Google's search engine, the first thing the auto-complete suggestion offers is 'salary', followed by 'age'.

Some people seem to think the McHughs own the Ploughing, because of the involvement of myself and Anna Marie. We always say: 'Look, our accounts are sent in every year and they are there for everyone to see.' But they particularly harp on about the salaries.

The people who obsess about my salary might not like to hear me saying that money, while it's grand to have, was never an issue for me. If it was, I would have moved on to something more lucrative years ago. My salary was never a bone of contention and I never once asked for a pay increase. It's only in very recent years that my salary has become a six-figure sum, yet other chief executives in the private sector are earning twice and three times this salary. That honestly doesn't bother me. I was, and still am, passionate about the work and my motto is to make every event a success. And if our finance committee ever wanted to give me an increase, well then, good and well. I think they pay me very fairly for my work and I certainly don't think they pay me less because I am a woman. The Association has always treated me very well, never questioning if I take a day off, or asking about my work output. As long as the job is done, they are happy.

My disinterest in the financial side of it is highlighted by the fact that I worked from 1951 to 1991 with no pension at all. It was actually our chairman who proposed at a meeting that there should be a pension for me. In less than a week, he had me down in Kilkenny, in front of an actuary who advised me on a pension. In all those forty years, it honestly wasn't something I had thought about. There was never a question of a pension. I suppose it was lack of knowledge on my part. Nowadays, I would be very concerned about our own employees looking after their pensions, but who thinks about pensions when they are young?

People are amazed to hear that I don't have a company car, nor do I want one. Yet, whenever I get a new car, I get the comment: 'Sure, it's well for you. Didn't you make millions at the Ploughing this year?' You have to take it on the chin, I know, but sometimes it's hard to take when you know the reality.

I don't like the focus on our earnings because the Ploughing is a private limited company with a board of directors made up of democratically elected representatives from each county, and an executive committee that reports to the board four times a year. Anything the NPA has, the Association has earned through the success of the event. Why should anyone not involved in the organization worry about it? If we were in debt, nobody would want us. I am only a custodian of the National Ploughing Association and it's my job to ensure that the funds are managed in the most efficient way possible. I have known hard times in the NPA when the money was not there, so that has made me very conscientious about minding the coffers. In the late 1950s I had to get a loan of £500 just to keep us going. You don't forget those days.

I mind the NPA's money like it's my own – maybe I'm even more careful with its money than I am with my own. I am meticulous about the paper trail for every penny the NPA spends. We don't pay by cash and we don't use casual labour. Our directors and volunteers are not paid but get travelling expenses. If the *RTÉ Investigates* team ever wants to come and have a look at how we run our affairs, we would throw the doors open to them. Although it would make for a very boring programme.

I was asked to comment when the IFA was going through that furore about its salaries in 2015. There were ructions over the pay and pension package of its general secretary, and the president of the organization, Eddie Downey, resigned over the issue. When I was asked for a comment, I thought to myself: How could I criticize another rural organization? And them in turmoil? I just couldn't do it. Everyone has to sort out their own difficulties. It was sad what happened in the IFA and I felt very bad for Eddie Downey, who was very unlucky with his timing as president. That problem was not of his making – it had been building up for years – but he was unfortunate enough to be the man at the helm when it blew up. He didn't get a chance to do all the things he wanted to do and it wasn't a nice way to go. I'd say it was an awful time for him and his family. It will take a long time for the organization to get over it, but it will because it's strong and capable. I firmly believe we need the strong voice of the IFA out in Brussels, arguing our case. We are a small country and Irish farmers need help to get their message across.

Apart from two once-off grants to run the World Ploughing Contests here, the National Ploughing Association gets absolutely no government grants, and not a penny from Fáilte Ireland, even though we bring so many people from abroad every year. We've even had exhibitors from as far away as New Zealand at the Ploughing. Last year we had people from more than fifteen countries at the event. Enterprise Ireland has been a great help in bringing international buyers over to the Ploughing in recent years, and that gives our exhibitors a huge boost. It's especially valuable to the companies displaying their wares in the innovation arena.

As I said earlier, you can't get a bed for miles around when the Ploughing is on. In fact, before we announce the venue, a year in advance, we have to ensure that our hotel rooms are booked first or we could end up in a tent. Now the press would have a field day if that happened.

NPA founder JJ Bergin said he hoped he would be leaving behind a working machine when he went from the Ploughing. I know I will be leaving behind a well-oiled machine, but I would like my legacy to be that the NPA has enough to stay in existence for many years to come. We have made investments that would get us through two bad years of weather and I think, at a minimum, it's where we must be. What kind of managing director would I be if I did not ensure that the NPA had enough funds to keep the event going for a few years?

We are so vulnerable to the weather. A classic example would be the Wednesday of the Ploughing in 2016. The heavens opened in the afternoon and we had torrential rain. Some people were calling it the million-dollar shower because it cost so much to put right. We spent almost half-a-million euro on repair works to ensure the site was ready for Thursday morning.

Our financial position allowed us to provide scholarships for two Kenyans to attend the prestigious Baraka Agricultural College in Kenya, to study sustainable agriculture and rural development. That was a one-off event, and it came about after Viva – the Volunteers in Irish Veterinary Assistance charity – approached us. They support livestock farmers in the developing world. We felt it was something that would fit in with our work. I had been in Kenya before, of course, at the World Ploughing Contest in Nakuru in 1995, and I remember how hard it was for them to till the land with their wooden ploughs. I was delighted the Association agreed to fund the scholarships because it was a concrete way of helping people. Now those two men have returned home to share the knowledge with their local community.

We also allow a certain number of charities to fund-raise at the Ploughing every year, and charitable groups get favourable terms when it comes to exhibiting. And we'd locate them, if at all possible, close to something that would draw crowds to that area. While we don't sponsor charities per se, it's nice to be able to give genuine causes a forum at the Ploughing.

In recent years, I have been asked, more and more, to speak at conferences and meetings about the success of the National Ploughing Championships and how I've managed, as a woman in business. I don't know if I do anything differently as a managing director because I am a woman. I am a great believer in bringing people along with you, rather than dictating to them from on high, but I don't think that's a trait specific to women. I treat everyone the same and give them as much responsibility as I think they can handle. In my experience, people always rise to the challenge. So, of course, I'm also a great believer in delegating as much as you can. No one person, no matter how brilliant they are, can run the whole show. It's not good for the organization, and it's certainly not good for the person who tries to do that.

To be honest, I think I got away with a lot more because of my gender. Yes, I had to fight my own battles to get things done, but I think men might have backed down a bit quicker than if they were facing another man. At the end of the day, what does it matter if you are male or female, as long as you can do the job? And women have proven that again and again. Attitudes have changed, too, and there would be uproar now if a raft of appointments were made and they didn't include women. Companies and politicians don't want to be seen to be giving all their plum jobs to women because it's just not good for their image.

Being a woman in a role like mine has brought lots of awards over the years, too. I certainly never expected to be recognized as businesswoman of the year in the Veuve Clicquot awards in 2013. That was particularly interesting because I don't drink alcohol. People who like the bubbly stuff will know that Veuve Clicquot is a premium champagne. The company started the awards for entrepreneurial women to honour Madame Clicquot, who took over the champagne house in the early 1800s after her husband died suddenly. She was only twenty-eight, and this was a time when women had no role in the business world, but she emerged as a formidable businesswoman. She had a huge impact on the champagne industry and pioneered significant change. She's still known as the grand dame of champagne.

As a non-drinker, I knew none of this when I got the call from Veuve Clicquot. I had a dilemma: should I tell them I didn't drink, or not? I decided to be upfront in case there was a champagne toast and I found myself looking around for a plant to tip the drink into. So, I met them in Naas and told them I didn't take a drink. I said I wouldn't be a bit offended if they wanted to change their minds and give the award to someone else. Fair play to them, they said there was absolutely no problem and they went ahead with the award and it was a lovely occasion in the French embassy in Dublin.

The name Veuve Clicquot probably rolls off some people's tongues, but I had to practise pronouncing it a few times to make sure I got it right. Then they took me, and the eight other award-winners from around the world, to their vineyard in Reims, where they planted individual vines in our honour, with little name plaques on them. I saw the vines of previous Irish recipients, such as publisher Norah Casey, designer Louise Kennedy and chef Darina Allen. I was in very good company indeed.

It's nice to think that the vine will be there for years, and perhaps some day the grandchildren will be in France and will decide to look up Nana's vine in Reims.

11. The highs and the lows

It was Monday, 19 February 2001. It was such an unremark-able day that I have no recollection of what I was doing at that time. Judging by the time of year, I would probably have been preparing exhibitor packs for the Ploughing, which was due to be held more than seven months later. I would have been going about my business as normal, with no idea of the turmoil that was about to hit the country. It was only afterwards that we realized the significance of the day. It was only afterwards that we heard about the worker in Cheale Meats abattoir, in Essex, who was feeding pigs that day and noticed that some of them were limping. The vet was asked to take a closer look at the animals. That was when the dreaded words were mentioned. Foot and Mouth. No one knew it at the time, but by then about sixty farms in England were showing symptoms of the disease.

Foot and Mouth is an extremely contagious disease that affects cattle, sheep, pigs and goats and if it took hold in Ire-land, it would wipe out our national herd and ruin the farming industry. News that it could be in Britain caused shockwaves here. We didn't have long to wait to find out. The following day, the British Ministry for Agriculture, Fisheries and Food confirmed that it was dealing with a Foot and Mouth outbreak.

I'll never forget the sense of dread when we heard the news. It's hard to put into words how anxious everyone became. We had all been talking quietly about it and praying that it wasn't Foot and Mouth. Now our worst fears were confirmed. You see, animals were travelling back and forth to Britain all the time and we knew that if it was in Britain, then it could spread to Ireland very quickly. It then emerged that, on the very day the abattoir worker in Essex had raised concerns about the pigs, a lorry had carried almost 300 sheep from Carlisle mart, in Scotland, to Northern Ireland. The mart had been contaminated with Foot and Mouth disease and twenty-one infected sheep on that lorry caused an outbreak on a farm in Meigh, County Armagh. Our hearts sank on 28 February when the Northern Ireland Minister for Agriculture, Bríd Rodgers, said animals had been slaughtered on the farm in Meigh and she believed Foot and Mouth Disease was in Northern Ireland.

It was an awful time, and I soon dreaded turning on the news. We were watching pyres of burning carcasses on British farms, on the television every night. There were so many animals to be slaughtered over there that the army was called in to help.

But we still recited our novenas, hoping against hope that it wouldn't spread across the Border. Then, on Thursday, 22 March, the then taoiseach, Bertie Ahern, stood up in Dáil Éireann and said an outbreak had been confirmed on a sheep farm in Proleek, near Jenkinstown in County Louth. Minister for Agriculture Joe Walsh had already introduced restrictions and they were immediately stepped up in a bid to stop the spread of the disease.

It was a case of all hands on deck then, and I must say it was remarkable the way the whole country pulled together to stop Foot and Mouth from taking hold. Even before the disease was confirmed here, people were doing everything they could to keep the disease out. Dublin Zoo closed its gates, St Patrick's Day parades were postponed, horse and greyhound racing stopped and golf courses closed. Even the Fine Gael Árd Fhéis was cancelled. You couldn't move without tripping over a disinfectant mat. It was amazing and so heartening to see such support for rural Ireland. The precautions worked and gradually the restrictions were eased back, so we were hoping that Foot and Mouth would be a distant memory by the time the National Ploughing Championships got underway in early October.

Things were looking good in late July so we had a press launch for the event, which was to be held in Ballacolla, again on the farm of David Lalor where we had previously been in 1995. The thinking was that the Ploughing would be an important stepping-stone on the road back to normality for Irish farming. Of course, we were in close contact with the expert group that was advising Minister Joe Walsh on Foot and Mouth, but we were optimistic that the worst was behind us.

But while things were going well in Ireland, the disease wasn't going away in the UK. As the event drew closer, we were getting the feeling from some exhibitors that they were thinking twice about coming. They were saying they would feel better and safer if it was cancelled. I suppose about twenty exhibitors said they were not prepared to come.

But while some people were concerned about the Ploughing going ahead, other exhibitors were telling us that they really needed the event to keep them in business. They had sustained serious losses because of the cancellation of so many other events, and because of the general uncertainty in the farming community. When farmers don't know what's ahead, they don't invest in new machinery or equipment, and this uncertainty had been going on since February. The

exhibitors desperately needed to get out and meet their customers and get some normality back into their businesses.

We were liaising closely with the Department of Agriculture throughout the crisis and both organizations shared the
view that the Ploughing should be a shining example of how
to prevent the spread of disease. People often give out about
the public service, but the Department of Agriculture was
absolutely fantastic during that crisis. They gave such good
guidance to us at every juncture. We had troughs and disinfectant lined up. We were taking every precaution and we felt
it was possible to hold the event while posing no risk to the
farming industry. In fact, we believed the Ploughing was
going to be a showcase of good practice in disinfectant and
risk management.

But farmers were very nervous about the disease. If you
saw a cow that was a little bit off-colour, you were afraid it
was going to happen to you. I remember one Englishwoman
talking about her experience when her farm went down with
Foot and Mouth. She said she'd never forget the sound of
silence when she went out into the field because the whole
herd had been slaughtered. It would send a chill down the
spine of any farmer. I remember seeing another farmer crying as he was interviewed, saying there would not be a sheep
left in Ireland if the culling continued.

About three weeks before the event, the NPA president,
Tom McDonnell, died in a farm accident with a combine-
harvester. Like many farmers in these cases, Tom was the
last one you'd expect to be involved in a farm accident
because he was one of the most careful farmers I knew. It
was an awful tragedy and coming at a time of such upheaval
with Foot and Mouth, it was a huge blow to the NPA. I well
remember the day of his funeral in County Louth – it

was Monday, 10 September 2001. There was a huge ploughing representation at the funeral and the words on everyone's lips were 'Foot and Mouth', second only to the grief of such a tragic loss. There had been several new cases in the UK on the previous day and word of another threat in Ireland had spread – the first such threat in a number of weeks. The writing was on the wall for the Ploughing.

At the funeral, I remember sitting in a car with the then president of the IFA, Tom Parlon, and a few others, and we all were very sombre about the likelihood of the Ploughing going ahead. Later that day our executive concluded that the National Ploughing Championships could not proceed, in the interest of the safety of the Irish farming economy.

So while the National Ploughing Championships continued throughout the war years, the recessions and every kind of weather you could think of, Foot and Mouth Disease stopped it dead in its tracks. That was the only cancellation of the Ploughing Championships in their seventy-year history, and it's a decision we never regretted.

We certainly did not relish telling exhibitors that the event would not go ahead, but our first priority had to lie in the safeguarding of Irish farming. We'd have no industry at all if Foot and Mouth took hold. It was a huge decision to cancel it, and we were very saddened about it, but if Foot and Mouth had spread because we went ahead with the event, well that would have been the end of the National Ploughing Championships. We would never have been forgiven. We would have had to close our doors, no doubt about that. How could you overcome that?

Early the next morning I went to the site in Ballacolla to do an interview with RTÉ's midland correspondent Ciarán Mullooly and RTÉ Radio's *Countrywide* broadcaster Damien

O'Reilly. After the interview, Ciarán noticed several missed calls on his phone. It was 11 September and two planes had just been flown into the Twin Towers in New York. That put everything into perspective, let me tell you. Our woes didn't seem that bad when we thought about the people on those airplanes, the people in the World Trade Center and all the families waiting to hear from them.

After our announcement, we thought we might be criticized for cancelling the event, or questioned on why we didn't act earlier. I was ready to explain it all, but to be honest, I never got a chance. The 9/11 attacks took over the entire news schedule for weeks and Foot and Mouth completely faded into the background.

There was a suspected Foot and Mouth case around that time in Cork, but it turned out to be nothing. But it occurred to me that if we had gone ahead with the Ploughing and the suspect case had become a reality, we could have been blamed for bringing it into the country. Then, just over a week later, on 19 September, the Republic of Ireland was declared Foot and Mouth free by the OIE, which is the World Organization for Animal Health. But once we had made our decision, there was no going back, and we were content that we had done the right thing.

Still, when the day came that the Ploughing was supposed to start, it was tough. Really and truly, it was like someone had died in the house. We were terribly lonely. We even got a few bouquets of flowers from exhibitors, and phone calls from friends of the Ploughing, which was very thoughtful. I kept looking at the clock, thinking I'd be doing such-and-such if the Ploughing was on.

We had spent quite a bit of money in preparation on things like the steel trackway, and I have to say that the company was exceptionally good to us and understood the position. In fact, all the contractors we were working with were very understanding because they also lost money on the event.

The host farmer, David Lalor, told us the land in Balla-colla would always be there and would be there for us the following year, if we wanted it. We had already settled on the venue for 2002 so we went to that landowner – Mark Moore, in Ballinabrackey, County Meath – and he said he'd keep the site for us until 2003. We were finding a way out of our problem and things were looking up again. We were incredibly lucky to escape Foot and Mouth that time. It was a real example of the whole country pulling together, north and south, rural and urban. Joe Walsh, who sadly passed away three years ago, aged just seventy-one, will always be remembered for his assured handling of the crisis.

Looking back now, it's amazing to think that the Second World War didn't stop the National Ploughing Champion-ships. The only change made during those years was that it was split into two venues in 1940 and 1941, with the Junior Classes being held in one county, and the Senior Classes going ahead in another. I haven't been able to establish why they split the event between two places. Could it be that they were afraid that the German fighter planes would spot a large crowd gathered in one area and, thinking they were flying over Britain, drop a bomb on the crowd? Perhaps by splitting the classes, they were splitting the crowd into half and therefore it wouldn't be as noticeable from the air? Or, indeed, it could have merely been a coincidence that the split in ploughing venues coincided with the war years. Maybe they were experimenting to see if it would work better with Junior and Senior classes being separated? Sadly, the people who were involved in running the event then are no longer with us, so I've been unable to find out why, and we have no Ploughing records that explain the move.

Of course, it would take more than a war across the water to stop farmers in their tracks. The RDS Spring Show in Dublin went ahead on Tuesday, 25 April 1916, the day after the Easter Rising started. Apparently, farmers showed 912 cattle at the event, but many of them were already in situ at the RDS before the fighting began. Not surprisingly, the show recorded a loss of £1,350 that year.

We've had to contend with absolutely every type of weather since the Ploughing started in 1931. I remember one lovely woman coming up to me when the Ploughing was in Meath in 2003. She said she had put out the Child of Prague on her windowsill to guarantee good weather, and she had also dug a hole on the site of the Ploughing and put another Child of Prague statue in it, for extra reassurance. I suppose she couldn't find a windowsill at the Ploughing. And we did have good weather that year.

The most unexpected weather came in November 1965, the morning the Championships were about to begin in Enniskerry, County Wicklow. I remember a knock on the door at 6.00 am. Nowadays I'd be awake from 4.00 am on the morning of the Ploughing, but I wasn't the managing director back then! It was Charlie Keegan at the door, with the astonishing news that the site was covered in snow. Now that was a first for us. You'll remember Charlie Keegan, the Enniskerry man who was the Republic of Ireland's first World Ploughing Contest winner in 1964.

Met Éireann's weather forecasters of the 1960s didn't have the technology at their fingertips that today's meteorologists have, so snow was something we did not anticipate. What could we do? We couldn't have people slipping and sliding on the roads going up to Enniskerry and sure, how could the ploughing go ahead in the snow? We couldn't do a thing except postpone the event until the following week. It was quite a disturbance for people because many of them were already in their cars and on the way. Hotels had been booked, so we did what we could to help with the competitors' costs when they had to return the following week. It was probably the most beautiful site we had ever had, with the hills and the view of the Sugar Loaf, but of course it would be out of the question now because of the road network and the amount of land we require.

One of our most disappointing turnouts was when the event was held in Finglas, in 1971. Back then, Finglas was in the countryside, but it was still seen as being in Dublin and I think the country people were afraid of the Dublin traffic. We couldn't rely on the Dublin people to turn out in force because there was not much to see except ploughing in 1971. I don't know what the turnout was, but the minutes book noted that it was very disappointing and it said we had incurred a financial loss. I've been searching the minutes books and I've found nothing to contradict my view that Finglas was the only year when the Ploughing suffered a loss. Finglas wasn't the only Dublin location for the Ploughing. It was held in Clondalkin in 1933. Of course, like Finglas, Clondalkin was a country village at that time. Cloghran in north County Dublin hosted the ploughing in 1942, and nearby Balbriggan hosted it in 1946. At the other end of the scale, I think Ardfert in County Kerry, in 1984, was one of our most remote locations. I remember people saying it was one of the best years in terms of sheer fun. I think we almost felt we were in another country, we were so far away.

Unfortunately for us, there was plenty of drama in Ard-fert. We announced the wrong ploughing result on the last day. I still cringe when I think of it. I didn't know anything about it until I got a knock on my bedroom door at about midnight. The mistake had been discovered and we had to act immediately. Naturally there was a lot of annoyance about it, but, with the goodwill of the competitors involved, we sorted it out satisfactorily in the end. (More about this incident in chapter 16, which is about lessons I've learned from my years running the Ploughing.)

New Ross in 2012 was another challenging event for us. We had a great site on the farms of David and Louise O'Dwyer and Peter and Tessie Kehoe. David's mother, the late Bridget O'Dwyer, was Queen of the Plough in 1965, 1966 and 1967, and his grandfather was a former chairman of the NPA, David O'Connor. You may remember David was the man who walked seven miles, with blistered heels, to the Plough-ing in Carlow in 1938 and still won the Championship.

The problem with New Ross was the traffic. It was a pity because ploughing is such a religion in Wexford and the peo-ple down there did everything they could to make it a success. We had a few wet days and the traffic didn't flow as we would have wished. One of the problems was that some people didn't follow the directional road signs on some routes. They probably thought they would be quicker following their sat-navs, but the sat-navs didn't know the Ploughing was on and directed them into New Ross and across the bridge, causing gridlock. It was one of the worst years for traffic, but the land was the best we could have wished for, considering the weather. No other land would have stood up to the rain so well. Despite the challenges, the ploughing public still turned out in their thousands, and while they gave us a slap on the wrist for the traffic problems, they still returned loyally the following year.

The weather may have been bad in New Ross, but not as bad as it was in 1981 when we had to contend with a full-scale storm in Wellington Bridge, County Wexford. The entire country was devastated by that storm, with winds of up to 80 miles per hour, floods of 2-feet deep in Monaghan and 30-foot waves bashing Tory Island. Our stands were flimsier back in 1981, so the wind tore the marquees, uprooted poles and shredded flags. We were lucky that no one was seriously injured by the flying debris. It was a struggle to avoid being blown off your feet, but the ploughing competitors soldiered on and we didn't have to cancel any competitions.

The weather certainly keeps us on our toes, and it never fails to amaze me how people still come to the Ploughing when it's raining and there's plenty of hardship. If it's very dry during the Ploughing, people would nearly feel cheated. They wouldn't get the use out of their wellingtons and rain gear. They'd say: 'Ah sure, it wasn't the Ploughing at all without the rain.' But it wasn't only rain that caused problems. We had one remarkable year in Urlingford, County Kilkenny, in 1986, when we actually had to dampen the ground, it was so dry. For the first time in our history, we found ourselves spraying internal roads to keep down the dust. I remember exhibitors having difficulty keeping their machines clean because of the dust.

It's not only the Irish weather that upsets our ploughing folk. The eruption of the unpronounceable Eyjafjallajökull volcano in Iceland caused another headache for us in April 2010. You'll remember it caused an ash cloud that grounded flights for days on end across Europe. The World Ploughing Contest was going on in New Zealand at the time and our competitors and officials were stuck in New Zealand for an extra two weeks. I didn't travel on that occasion, and I remember one of their family members ringing me and asking: 'Is there anything at all you can do to get them home? There's so much work to be done on the farm.' And I said: 'I can hardly get up in the air and clear the cloud away or fly them home single-handedly.'

And then there was the outbreak of SARS (severe acute respiratory syndrome) in Toronto in 2003, which caused more chaos and worry. The World Ploughing Contest was being held in Guelph, Ontario, that year and the Irish team was flying into Toronto, which had experienced two waves of the SARS outbreak between February and June. The World Health Organization had removed Toronto from its list of SARS-affected cities in July, and the ploughing contest was being held in August. There was still a lot of worry in people's minds because SARS was in the news globally. In fact, the forty-fourth SARS death in the Toronto area happened two days before the World Ploughing Contest got underway. However, it emerged that the man had been fighting the illness for months.

Naturally, we were all wondering if it was safe to go. I wasn't worrying so much about myself, but I would hate to think that I was putting a competitor's health, or the health of their family, in jeopardy because of the trip. And of course, there was the worry that we might bring the disease home with us. But we kept abreast of the developments and the advice from the World Health Organization and we decided it was safe to travel. The World Ploughing Contest went ahead as planned and, more importantly, everyone returned home in robust good health, including myself.

I think the Nissan Classic cycling race was a major

influence when it came to getting non-farmers to come to the Ploughing. We landed a tremendous coup in 1989 when we persuaded the organizers of the cycling race to pass through the Ploughing site, which that year was in Oak Park, County Carlow. Because it was in the Teagasc tillage research centre, we had internal roadways going through it, so the cyclists could enter one way and leave another way. I think it was one of the greatest things we'd ever had. Most people would never have got a chance to see the Nissan Classic and here it was, going through our Ploughing site. The speed of the cyclists when they went through the site was fantastic. I couldn't overestimate the value of that to the Ploughing. People came from urban areas to see the race and I believe they never stopped coming to the Ploughing afterwards.

I remember we had a funny incident in Tullow, in 1967, when a group of Cork competitors couldn't find their digs. They were walking up and down the street with their wellingtons in their hands when the rest of the town was asleep. Anyway, the gardaí came upon them and discreetly noticed that the wellingtons were holding bottles of poitín. It was late at night and everyone was tired, so the gardaí agreed between themselves that it was easier to show them to their digs than arrest them for the poitín.

As you can imagine, I have to be prepared to always expect the unexpected in this job. I remember we had a huge crowd gathered for the press launch of the Ploughing one year and we were about to serve them lunch when the caterer came out and said: 'We have no electricity. You'd better delay them with speeches until it comes back on.' And we covered it up well and no one at the tables knew there was a crisis going on behind the curtain.

I remember another occasion when we had announced a

venue for the following year and we got a phonecall from the National Roads Authority. They feared that the venue was interfering with a new road being built. And indeed it was. The road would have cut the site in two and work would have started before the Ploughing. Luckily, we got an alternative venue without too much trouble.

There was another year when we had announced a venue and even gone as far as sowing grass seed. We got a phone call from the county council, asking to speak to us. They told us we'd have to seek another venue because the roads were not adequate. I remember Anna Marie telling them: 'You are too late. We're going ahead.' And we had three very successful years in that venue. Once the council knew it was going ahead, they did everything they could to assist us.

You would think that I've come up against every obstacle at this stage, but every year something new crops up and you have to deal with it. There was the year when it was decided by the Health and Safety Authority that we should be recognized as a construction site during the periods when we are preparing the site, and dismantling everything afterwards. Ever since, we've had to ring-fence the entire venue. We also have to present an event management plan detailing every single happening over the three days. If anything goes ahead that's not included in the plan, we are in trouble, and the buck stops with me.

We always comply with requests from the regulatory agencies, but there was one exception. That was when I was told by a fire officer to have a morgue on the Ploughing site. A morgue! Well, I thought it was the weirdest thing I'd ever heard of. He said it was a precaution, in case someone died on the site. I thought things had gone too far at that point.

Where do you draw the line? We have a doctor on site, the

Order of Malta, several first-aid centres and a large medical centre. Really and truly, could we do any more, apart from opening a fully staffed hospital on every Ploughing site? You could tie yourself up in knots trying to accede to requests like that, so I just refused to agree to it.

But I find that in the majority of cases, the emergency services and statutory regulators really do want to work with us for the good of the event and the public. Either myself or Anna Marie will always attend the services meetings. We would never dream of handing this responsibility over to an agent because if they are prepared to send their senior people to meet us, then we owe it to them to make the time. It's very beneficial in the long run because we all know each other well by the time the Ploughing comes around.

I would also bet that no one would think that we have to get permission from the Irish Aviation Authority (IAA) to fly balloons on the site. The big blimps that help people find their way to the NPA headquarters, or to any other exhibitor, must be licensed by the IAA before they can be flown. It's in the event of a blimp breaking loose and going up in the air. I suppose it could pose a safety risk to an airplane because they are very big balloons. Our safety officer also has to inspect them to ensure they are tethered down correctly. When the recession took hold, the number of balloons went down, but now they are flying high again all around the site. We also have to get permission from the Irish Music Rights Organisation (IMRO) to play music on the site. These are things that the public never thinks of, but we have to be on the ball.

And who would believe that the National Ploughing Association has a herd number? We have to have the herd number to cater for the livestock coming onto the site. All livestock are kept in a holding pen until they are inspected by Department of Agriculture officials. Then they are released to take their place at the livestock exhibit.

We get as near as possible to having everything perfect every year, but some new quandary will always present itself. Two relatively new environmental farming schemes – Glas, and the greening scheme – have thrown up problems because they tie up land for farmers and make it difficult for us to secure sites as a result. So we now have to prepare arguments to present to the European Commission in Brussels so that farmers participating in these schemes can get dispensation status to allow us to hold our ploughing competitions. I sometimes worry that we will become such a regulated farming community that the humble plough will become redundant.

There is lots of talk now about 'min till', or minimum tillage, which involves preparing the soil without ploughing. There are arguments that it locks in carbon in the soil and so combats climate change, but it's hard to find research that proves that it achieves the same results as ploughing. The plough has served us well down through the years, and I hope we will not see its demise anytime soon. Many sheds around the country house a match plough that may not have been used for ten or fifteen years, but families are reluctant to sell on the heirloom because it represents the sport that someone loved and excelled at. The beauty of the plough is that it does not really age or date, so a son, daughter or a grandchild can always decide to rekindle the love of ploughing.

It's amazing what some people will try, given that the site is so well policed. There was one year when we had to have a stand removed because an observant steward discovered that the standholder was selling illegal drugs. The next morning, I brought a garda to the stand before 7.00 am. He told me he'd deal with it, and the squad car arrived shortly afterwards. The information was correct and the man, and his stand, were removed before the crowds started to pour in. That was the only time we had to get a stand removed.

And then sometimes people expect trouble when there's none at all. I remember we had a group of Travellers living close to the entrance of the site at Vinegar Hill in Ferns, in 1998. People said, quite unfairly, that they would steal everything from the site. I don't think it's fair to prejudge people and I didn't know these people, so I decided to go down and meet them. I explained how we were having the event and they said: 'We'll mind the site for you, Ma'am.' And they never touched a thing on the site. But I remember late one evening we spotted someone in the field loading up equipment and it looked a bit suspicious. We discovered that it was a farmer who was taking the equipment and he said it had been on the site a few days and no one wanted it. Now if we hadn't spotted that, and the equipment had disappeared, everyone would have blamed the Travellers for it.

Every year we have a running battle of wits with the three-card-trick men who try to sneak into the Ploughing, even though they are not allowed. As soon as they are spotted, they are ejected. They travel the country's shows and fairs with their little fold-up tables, and we've come to recognize many of the faces at this stage. It's often teenagers who get sucked into their tricks and it's not fair on the vulnerable young people who could lose all their money. We've never allowed gambling at the event because we don't want people leaving the Ploughing with a bad taste in their mouths.

On another occasion my husband, John, spotted this man

selling sunglasses or something similar on an internal road-way at the Ploughing in Oak Park, County Carlow. John knew we would never have allowed someone to set up on the side of the road because of the traffic obstruction, so he told the trader he shouldn't be there.

'Oh, I was talking to Mrs McHugh and she gave me this spot,' the man told him confidently.

'She didn't,' John shot back at him.

'You don't know Mrs McHugh,' the man said, giving John his chance for the perfect comeback.

'Well, I ought to know her because I slept with her last night,' said John.

And the man had no answer to that. Lord, I roared laughing when I heard that.

Another day I was taking my daily trip to the ploughing fields when I spotted a man selling sweets out of the boot of his car. I went over and asked him did he have a licence to do so. He said he had approval from the office, but his confidence quickly crumbled when he heard who I was. Of course he didn't have approval, but sometimes you have to have a bit of give and take. We agreed a price for the shop there and then, and that man is with us since.

We always get our fair share of chancers at the Ploughing. Every year we get people trying to get in without paying. One year in particular, a ticket-seller got an awful slew of people coming, one after the other, all looking to get in for free. They were saying they were related to the site owner or they knew so-and-so and they shouldn't have to pay. He was at the end of his tether when this nice big car approached him and the driver said: 'I have the Turkish ambassador in the back of the car.' The exasperated ticket-seller said: 'I don't care who you have in the back of the car, you're going to have to pay to get in.' And the Turkish ambassador did pay, just like everyone else.

There was another year when the taoiseach was expected – I think it was Bertie Ahern at the time – and a Garda motorbike made his way down through the long lines of traffic to escort him into the site. The Garda spotted a black Mercedes and presumed it was the taoiseach, so he beckoned to the driver to move out of the long queue of traffic and follow him. They flew by all the cars with ease, and straight into the site. When they arrived in, didn't the Garda realize that there was no taoiseach in the car at all. It was just an ordinary visitor. The driver was no doubt delighted to have skipped the queue of traffic, and who'd blame him?

But the nice thing about the celebrities coming down to the grass roots of Ireland is that they can leave their airs and graces behind them, if they have any, and in fairness most don't. We often get calls to say that this person or that person was thinking of going to the Ploughing and what plans would we make for their visit. I take pleasure in saying that they are most welcome and we can send them an information pack, but they will have to get a ticket and walk in like everyone else. I've never expected to be treated as anyone special and I think it's important that we keep things simple at the Ploughing. The ordinary person walks the trackway with politicians, sporting heroes and celebrities and sure, we are all just the same at the end of the day. Although, some people do become overawed when they meet VIPs at the Ploughing. I'm thinking of the time I was about to introduce President Mary Robinson to a horse ploughman, and he took us both by surprise by suddenly taking off his cap and dropping down on one knee in total respect of her position.

I must admit I've been caught out a few times myself when

a VIP is brought to my office during the Ploughing for a meet-and-greet and I don't have a clue who they are. I am well used to covering my tracks when that happens and we have a lovely chat until they walk away and then I have to ask: 'Who was that?'

Nor do people always recognize me. On occasion, I have tried to access the ploughing site very early in the morning or late at night without an admission sticker and the security guy would just say to me: 'Sorry Ma'am, the boss says no entry after such a time or until such a time.' I've had to ring the security company a few times so that their guy would believe I was 'the boss', but these things are always taken in good spirit.

Of course, we all got up to a bit of devilment every now and then during ploughing competitions. Sometime in the early 1980s I was at a ploughing competition in Limerick and John Somers from Wexford was competing. He got a bronze medal in the 1980 World Ploughing. Anyway, John arrived in the field with his suitcase and left it on the headland. I got a nice fistful of clay and put it in his suitcase, closing it again carefully. But I didn't really think ahead because he was staying in a friend's house that night – a beautiful home – and when he opened the suitcase the clay went all over the bedroom carpet. I felt a bit bad about that!

Speaking of suitcases, I'm liable to come home with anything in my suitcase from the Ploughing. One year we arrived home with a pair of very large men's trousers. I still don't know how they got in there, but I suspect someone was helping me to pack and everything was flung into the bag at the last minute, including the Bed and Breakfast owner's trousers. Another year I arrived home and opened my suitcase to find a nice selection of men's shirts. A man in Kilkenny was probably opening his suitcase at the same time and wondering why Anna May's nightdress was in there. We had identical suitcases and they got mixed up as we were packing. It's a similar story when the container is being loaded for the return journey at the World Ploughing. Anything could end up there. Anna Marie is still trying to find the owner of a bag containing a selection of women's lingerie and a fine pair of hiking boots.

I called this chapter 'The highs and the lows', but when I think back on my years in the Ploughing, the highs definitely outnumber the lows. The Ploughing is about getting to the heart of things, which is what I love about it. Ploughing takes you back to nature, to the cultivation of the soil to feed the people. What could be more basic and natural? I remember being at a workshop for ploughing judges in March 2017 and the news was full of Brexit and Donald Trump and someone said: 'Look at us here, the whole world going mad around us and we're in a tent, in the middle of the field, talking about ploughing. And at the end of the day, we wouldn't have a bit on the table without the plough.' That's what it all boils down to.

12. Presidents and politicians

If you had told me when I started work in 1951 that one day I would be meeting world leaders and travelling the globe, I would have laughed at you. Back at the beginning, my main priority was packing JJ's little school case for the Ploughing and wondering if we'd be lucky enough to get a mention in the local paper.

But it's amazing how the Ploughing has opened the door to so many opportunities. Our involvement in the World Ploughing Organisation has taken me around the world more than a few times. And thanks to the growing importance of our own event, I have come face-to-face with our own political leaders and so many other well-known people in Ireland. I'm not inclined to get too excited about meeting people like that. I suppose I see it as an acknowledgement of how far the Ploughing has come, that so many well-known people want to be associated with it. However, a few of those encounters stand out in my memory.

I'll never forget meeting the French president, Charles de Gaulle, at the World Ploughing in 1961. It was being hosted in Grignon, just outside Paris, and the president hosted a reception for the participants. Like many French people, he was extremely stylish, a very well-dressed man. He was wearing these immaculate cream or white leather gloves and I remember being surprised that he didn't remove them before shaking hands with us. We were all struck by his height, and his charm. He spoke to us as a group, welcoming everyone and hoping that we would enjoy our time in his country. It was a big thrill to meet him. Charles de Gaulle had special time for the Irish, of course, because of his family roots in County Down. He spent six weeks in Kerry and Connemara after he resigned from office, in May 1969. He was a devout Catholic, by all accounts, and there's a lovely photograph of him and his wife, Yvonne, coming out of Sunday Mass in Cashel, Connemara. Kerry photographers Padraig and Joan Kennelly documented the whole trip. Poor de Gaulle died of a heart attack a year-and-a-half later.

Incidentally, de Gaulle twice blocked Britain's efforts to join what was then known as the European Economic Community (EEC), in 1963 and 1967. Maybe he knew more than we did and could foresee that the British relationship with the European Union would end in tears someday.

I don't mind saying that I went out of my way to avoid meeting Robert Mugabe. He was prime minister of Zimbabwe when the country hosted the World Ploughing Contest in 1983 and he's still ruling the country, as president, thirty-four years later. We were in Harare and he came to wish everyone well. He was accompanied by a motorcade of seven cars. All the competitors stood under their country flags, but I had to disappear to the back because I just couldn't shake his hand. It didn't feel right. I couldn't tolerate the wealth he had and the poverty that his people were living in. Someone pointed out his residence as we were driving by very early in the morning and there were already three or four men working around it. It just didn't seem right.

Zimbabwe will always stand out in my mind as being a very beautiful country, with lovely fuchsia everywhere. It was such a culture shock for us when we arrived. I suppose we white people are used to being in the majority wherever we go, but in Zimbabwe we were a few white faces in a sea of black. It was such an unusual feeling. And there was so much poverty there that we felt compelled to share whatever we had with the local people. I remember one occasion when I gave a little boy a big packet of biscuits. The poor child shared the biscuits to the extent that he didn't have a single one left for himself.

We were inundated with visits from Irish missionary priests and nuns when we were over there, so when we were leaving we left as much as we could with them. We gave them clothes, shoes, toiletries, anything that could be useful. People were very generous. We knew the missionaries would make sure that the neediest people got them.

The Zimbabweans' approach to time was quite different from ours. Some foreigners think Irish people are very laid back about starting events on time, but in Zimbabwe they were even more relaxed. Something might be due to start at 7.00 pm, but if you arrived at 8.30 pm it would only be getting underway. I try to be a very punctual person and I'm always keen to ensure that our ploughing meetings start on time. And if a ploughing competitor is late for a match, we do not wait. The competition starts on the dot, whether a competitor is stuck in a traffic jam or not. But it was a completely different matter in Zimbabwe. And I just had to accept it. That was the way they operated and I wasn't going to change them.

I remember being driven up a dirt road and we couldn't see the driver, there was so much dust in the vehicle. You would hardly see a vehicle in front of you with the dust and I was always afraid that we would have an accident. That reminds me of a former curate in our parish of Arles, Father Donal Deady, who was killed in similar circumstances when he was working in Nairobi. In December 1986 he was travelling with three other Irish missionary priests when there was a collision between their car and a bus. The tragedy was that he was due to return home, to take over a parish, a month later. We visited his grave in St Austin's Cemetery, Nairobi, and placed a little Irish flag on it, when we attended the World Ploughing Contest in Kenya in 1995. He is buried opposite another Irish person, the lay missionary Edel Quinn. Although she was very ill with TB, she spent the last years of her life in east Africa, spreading the message of the Legion of Mary.

It was such a wonderful opportunity to see Africa because I had never been anywhere like it before. The people were so pleasant and kind and everyone was trying to eke out a living, even down to the older women selling their crochet work on the side of the road. A little means so much to them.

Mass in Kenya was a very special experience. I remember going into the church before Mass began and the children were already singing hymns. They were all dressed so well for Mass, particularly the little girls. The music was amazing. It was so joyous and rhythmic. They would hit the heel of their hands off the seat in front of them and it created a marvellous drum-like sound. What really impressed me was the way they did the offerings. They came with eggs, bread, fruit, vegetables, even a live chicken. I'd say there were fourteen or sixteen people bringing up offerings. I remember wondering would the procession of gifts ever end. The Mass went on for hours, with singing and praying before the priest arrived and after he left. It was a wonderful thing to see.

But back to my meetings with the famous and the infamous. We have been very fortunate with the support we have received from our presidents and taoisigh. It all started with Éamon de Valera. He was taoiseach when he first visited the Ploughing in 1938 and that gave a great boost to the Ploughing. People started to think that this Ploughing must be a big thing if the taoiseach was going to it.

We never entertained Ireland's first president, Douglas Hyde, at the Ploughing, but there was huge excitement when his successor, President Seán T. O'Kelly, came to the Championships for the first time in 1946. I wasn't there, of course, but I've heard that he was a very stylish man and he brought great glamour to the event. There was a bit of drama surrounding his visit to the Ploughing in Bandon in 1950. The Wicklow man decided to tour the plots to see his fellow county-man, John Halpin, in action. He was going through a gap when one of his wellington boots got stuck in the muck. According to the NPA minutes, one man held the president up by the shoulder while another attendant retrieved the wellington. 'The little man from Roundwood proceeded on his tour as if nothing had happened,' JJ Bergin recorded in the minutes.

I remember then president Cearbhall Ó Dálaigh coming to the Ploughing in the mid-1970s and he told me in private conversation that he sometimes felt like a bird in a gilded cage. Of course, presidents didn't move out of the Áras much at all in those days. It only really started with Paddy Hillery, and then Mary Robinson took it to a whole new level. I really admire her because she truly broke the mould when she became the first female president. She acquitted herself extremely well, both here and abroad, and she has been a wonderful ambassador for Ireland. And look at the great work she's done since then, on human rights and climate change? We can be very proud of her.

Paddy Hillery, Mary Robinson, Mary McAleese and Michael D. Higgins have all made a point of coming to the Ploughing and have been very good to us. President Hillery always said he loved meeting my daughter, Anna Marie, because his daughter, Vivienne, was the same age. Sadly, Vivienne died after a long illness in 1987, just before she turned eighteen.

You get invited to all sorts of places in this job, so I've found myself at receptions in the US, German, French and Indian embassies for garden parties or other functions to mark special days in their countries. I suppose they try to invite people from every walk of life and I'm asked because of the Ploughing. They are always lovely occasions, not too formal, but you would definitely make an effort and dress up for the event. The embassies and the ambassadors' residences would leave you feeling that your own home was very modest indeed. I remember going to the ladies' powder room in the US Ambassador's residence and my God, it would take the sight out of your eye, it was so beautiful.

We have a great relationship with the French embassy and I was very honoured to receive the Ordre du Mérite Agricole, one of France's highest and oldest honours, created in 1883. It rewards people who render exceptional service to agriculture. The presentation was made to me by the French Ambassador to Ireland, Jean-Pierre Thébault, during a ceremony at the France Agriculture Pavilion at the 84th National Ploughing Championships in Ratheniska. Mr Thébault is a lovely man and very, very relaxed, as his driver could testify. I say that because the ambassador came down to my office in Laois to give me an overview of the award before the ceremony. We got on very well and it seemed that he had all the time in the world to chat, but you could tell the driver was getting impatient to get back to Dublin after about an hour. It was quite comical to watch him. He literally walked up and down outside the office window and at one point he actually came in and pointed to his watch, but we were having such a good chat that the ambassador left at his own pace.

If you were to ask me who made the most impact on me, of all the well-known people I've met, I think it would have to be Mary McAleese and her husband, Martin. She had such a relaxed way with her when she was president. I remember myself and John were invited to Áras an Uachtaráin for a Christmas dinner not long after she became president. There were about eighteen of us at the dinner – professors, heads of universities and the like – and I remember John saying to me beforehand: 'How on earth am I going to talk to those people? Can you not go on your own?'

As a quiet country farmer, he really dreaded the thought of going up there. I was a bit nervous too, to be honest, because I didn't know her very well at the time, but I didn't want to let on or it would make him worse.

So I said: 'John, the first thing you have to do is to go into Carlow and get a new suit.' And the reply was: 'Sure, what's wrong with the ones that are hanging in the wardrobe?' I'm sure a lot of women have heard that comment from their husbands at some stage! Needless to say, he went off to Carlow and got the new suit, whether he needed it or not.

Anyway, who was the last couple to leave the Áras that night, but us? Mary and Martin McAleese were so nice to us and she really put John at his ease. She just seemed to click with him. It must be the easy way she has with people. I remember she was showing us around the Áras and John pointed at this lovely gold raised wallpaper and said: 'Now, could you change that if you felt like it?' and she said: 'John, if I changed that wallpaper, I'd be ran out of the place.' They had a great understanding of each other and he always loved her after that.

I don't know how she heard, but the day John died she was on the phone to me to express her condolences. The RTÉ broadcaster Seán MacReamoinn died the same day as John. Incidentally, as a young man, Seán MacReamoinn covered the Ploughing for RTÉ. I suppose there's hardly a person in RTÉ who hasn't been dispatched to the Ploughing at some stage. Mary McAleese had to attend the broadcaster's funeral, but Martin McAleese still found time to come to John's funeral, which meant so much to the family. I thought it was such a nice gesture. I always thought that if Martin ran for president, he could have easily won the race, he was so well-thought-of and well-liked. Brian Cowen, who was Minister for Finance at the time, came to John's funeral too, which I thought was very kind. John wouldn't have been well-known in those circles. He always took a back seat, so it was nice that people honoured him.

After John died, Mary invited the family to the Áras for St Patrick's Day and she always kept in touch with us. I've heard similar stories from other people about her thoughtfulness and warmth. She came back to the Ploughing in 2012, the year after she finished her term of office, and it was lovely to see her again. I haven't met her since then, but I always love to read about what she's up to in the world of academia. Any university student would be lucky to have her as a lecturer.

I've been up in the Áras a few times since that first occasion with John. When the NPA was celebrating its eightieth year in 2011, Mary McAleese hosted an event to mark it. I'll always remember one of the lads coming out on the steps of the Áras and going back in and repeating this a few times. And I said to him: 'What are you at, at all?' and he said: 'Now I can tell people I've been in the Áras three times. Not many people can say that.'

I was delighted to be invited to Dublin Castle for Mary McAleese's second inauguration in 2004, and seven years later I was sorry to see her leave the public stage. I think she brought a great warmth to the office of the President. She was so down-to-earth and never put herself on a pedestal. She could speak to the highest-ranking individual and the child in the pram and make both feel special. We all felt that she was one of us, and I think that made Áras an Uachtaráin more accessible for everyone. One of her greatest legacies, undoubtedly, was bringing Queen Elizabeth to Ireland. I'm sure Mary McAleese was very proud that everything went off so well. Neither she nor the Queen put a foot wrong during that visit. You couldn't have dragged me away from the television when the royal visit was on. I was absolutely glued to it.

When Michael D. Higgins became president in 2011 and he accepted our invitation to come to the Ploughing in 2012, I said to myself: 'Lord, this man will know nothing about Ploughing at all. He's a poet and an academic.' I was a bit concerned about how well he would be received by the farming public, and how he would fit in at the Ploughing Championships, as opposed to a big art exhibition. I couldn't have been more wrong. He stood up at the Ploughing and amazed everyone with his speech. He is a native of County Clare, and he talked about spending his early years on a small farm in Newmarket-on-Fergus. He said a lot of his poetry was influenced by those farming memories. It was just wonderful to listen to him. As soon as he mentioned the farm in Clare, my concerns vanished. I knew he understood the rural culture and you could tell that he was really at home meeting the ploughing competitors and walking around the site.

I immediately took to Michael D. and his wife, Sabina, that day. They are both extremely warm and affectionate people. Of course, it wasn't his first time at the Ploughing. He came to the Ploughing in Athy the previous year, when he was on the campaign trail for the presidency, so he knew what to expect, but I wouldn't have had the same interaction with him then. Some people ask if we keep a stack of wellies for our VIPs in case they come unprepared. No, we don't. By now most people know what to expect when they turn up at the Ploughing, bar the odd unfortunate woman in high heels or delicate footwear. Michael D. and Sabina don't wear wellingtons, but they do come equipped with good strong shoes.

Everyone loves Michael D. at the Ploughing, but I've noticed he's especially loved by young people. You'd see them all running after him, calling his name and lining up to meet him.

We meet all sorts of people at the Ploughing, but I have to be very careful not to show any political bias in this job. You must be fair with everyone. I remember there was a ploughing match in Cork and the Fine Gael politician Mark Clinton was Minister for Agriculture at the time. This particular colleague in the Ploughing would never tell us what his political affiliation was, but when Mark Clinton arrived, the man nearly knocked everyone down to shake hands with him. I said: 'Ha, we know your political leanings now!', but he said he was just being friendly and respectful. No one believed him, of course. People can be very quiet about their political affiliations in the countryside, but some households were always known to be strong Fianna Fáil or strong Fine Gael houses and you'd presume that they would still cast their vote in that direction.

In my case, my father was a Fianna Fáil supporter so we followed suit when we were younger. My husband, John, was a Fine Gael man, but then when you get to know your local representatives you vote for the people you know, regardless of the party. I was never that strong on party affiliation.

Some politicians would practically knock you down at the Ploughing if they saw a TV camera approaching. If you were standing in front of them, they'd nearly put their two elbows up on your shoulders to be seen. There are times you'd love to say something, but at the same time it is great that they come because they are part of what makes it what it is. I don't know how genuinely interested some of them are in the Ploughing, but they'd always let on that they were anyway, and sure who'd blame them?

But we must be very careful not to allow them to use us to endorse their cause. I remember there was an election coming up and the taoiseach of the day was coming to do his usual press conference in the press room at our NPA headquarters. We discovered that he was going to be flanked by his party's candidate. We couldn't allow that to happen because people would think we were endorsing his candidate. We told his handler that, in no uncertain terms, and so they moved the press conference outside the headquarters. The taoiseach was still flanked by the candidate, but at least we were not facilitating it.

Similarly, we cannot allow any person or group to take ownership of the National Ploughing Championships when it suits them. We are a totally independent rural organization but believe me, in the past some leaders of various groups have tried to make it look like they are running things at the Ploughing. I remember one time a leader of an organization exhibiting at the Ploughing moving directly in front of me to welcome the President of Ireland to the Championships. He actually stood out in front of me. That agitated me, I can tell you. I had to say: 'Stand aside, please, this is our event. You can do your business later.'

And in the past we had to post two NPA people on both approaches to the bandstand to stop some politicians and, occasionally, organization leaders, from getting up on the stage and parking themselves in the front row during the opening ceremony. Although, a lot of that was about the personality of those men. Times have changed, though, and I couldn't see that happening now.

The 1997 presidential campaign was in full swing when the Ploughing was held in Birr, County Offaly, at the end of September and beginning of October. Election day was to be 30 October, so several of the candidates turned up to press the flesh.

There was a big field that year. As well as the eventual winner, the Fianna Fáil candidate Mary McAleese, we had Fine Gael politician Mary Banotti, Chernobyl campaigner Adi Roche, Eurovision Song Contest winner and later MEP, Dana Rosemary Scallon, and Derek Nally, a retired garda and victims' rights campaigner. It was the first time there were five candidates in a presidential campaign. That number of candidates was only matched, and indeed exceeded, in 2011 when seven candidates ran, and Michael D. Higgins came out on top that time. But it was the 1997 campaign that really stood out for me in Ploughing terms because there was such a melee around the candidates. I remember Mary Banotti saying she would have had to travel thousands of miles to make contact with the number of people she had met at the Ploughing. She had spent six hours canvassing on the day she visited and I was impressed by her. She was there at the same time as Mary McAleese and despite the size of the site, both candidates managed to collide outside our headquarters. It was a media scrum, but people were delighted to get a chance to see the candidates up close, so it added to the event. We set up a microphone in a central location for the candidates who wished to address the crowd and they were very well received.

But that was nothing compared to the melee around Bertie Ahern, who was a frequent visitor to the Ploughing when he was taoiseach. Even though he was a proud Dub, he was always a very popular visitor to the Ploughing. People just loved him. Like Mary McAleese, he had the common touch.

The 2006 Ploughing in Tullow is probably not such a happy memory for Bertie. That was when he was being asked questions about payments and 'dig-outs'. He had just done a big interview with Bryan Dobson, explaining the payments he got from friends after his marriage separation, on RTÉ's *Six One News* on Tuesday, 26 September. He lay low after the interview but was due to arrive at the Ploughing on that Friday, so the journalists were dying to get a hold of him. I remember them waiting for his helicopter to land and they were all ready to pounce on him. To be honest, I felt a bit sorry for him, arriving into that rumpus.

We were also hosting the World Ploughing Contest that year and we had arranged for him to meet the overseas competitors. Obviously that would have been a big thing for a ploughing competitor, to meet the prime minister of the country where they were competing. But there was a strong feeling from the media that he was avoiding the reporters and there was a lot of speculation about his visit. Would he even turn up? We had very little contact with his office in the days before his visit, but I never had any doubt. I knew he wouldn't let us down.

So he got out of the helicopter and into the jeep with me. Then his team said Bertie would not be going down to the ploughing plots because there was too much controversy going on and it would cause a lot of hassle trying to marshal him around the site with the media on his tail. They said he would just come into the NPA tent, show his face and go. I was sitting in the back of the jeep with Bertie and he turned to me and said: 'Where are we going, Anna May?' I said: 'We're going to the ploughing, Taoiseach.' And that's what happened. That was Bertie. He wouldn't be dictated to by his public relations people. He was the boss.

I remember another occasion when his handlers literally had him running around the Ploughing site. We went into a quiet corner of a stand and he turned around and said to his handlers: 'Now, you listen to me. I am walking around. I am not running. I am not afraid of the people of Ireland and I am meeting them.' And that's exactly what he did. Anna Marie has walked many taoisigh around the Ploughing and she says she never once heard a bad word directed against Bertie.

I liked him then and I like him still. He has plenty of charm. He's wearing his age extremely well. I think he would make a fine president if he decided to run. We were having this discussion recently and someone said he couldn't run because he had dirtied his bib. 'And couldn't we wash it?' I said jokingly.

There are some politicians who have been great friends of the Ploughing, people like Joe Walsh, may he rest in peace, and Ivan Yates, both former ministers for agriculture. They have always attended the Ploughing and their appearances never failed to generate great interest from other visitors. It adds to the day out when you see people like them out enjoying themselves at the Ploughing.

I have to say, I generally have a very good relationship

with the press. I might not always like what they write, but that's their business. As long as it's accurate, and fair, I have no issue with it. I've mentioned Seán MacConnell before and I couldn't forget people like Joe O'Brien, who covered the event for RTÉ so well for so many years, and Ray Ryan in *The Examiner*. Both retired now, both real decent people and both have returned to the Ploughing since. In fact, Ray still comes out of retirement every year to cover the Ploughing for *The Examiner*. Sure, he knows it inside out.

I think I'm always that bit more relaxed when I'm being interviewed by fellow Laois people, such as Eileen Dunne, Seán O'Rourke and Claire Byrne. And sure, we claim Miriam O'Callaghan as one of our own because her Mam was born in Ballylinan. It's nice – and only proper – to see the Laois talent rising to the top in RTÉ.

It's great when you meet people in the media who are prepared to give a balanced story and 'the real facts' – not just a colourful version. I think the agri sector is served well by the current crop of journalists in publications such as the *Irish Farmers' Journal*, *Farming Independent*, RTÉ, *The Irish Times* and *The Examiner*.

I'm well used to dealing with the media as the face of the NPA, which is no bother to me, but it's a very different thing when people just want me to be myself. I did the *Meaning of Life* with Gay Byrne in 2016, and I must admit that it frightened the life out of me beforehand. My first instinct was to say 'no' when I was asked, but Anna Marie told me: 'You have to do it. It's not everyone who gets a chance like that.' I think they were after me for six months to do that programme. My feeling was that I was not, I suppose, at a level to do that. I would never put myself in the same category as the other people who did the programme. People like Maeve Binchy,

Terry Wogan, Brian Cody, and Mary Robinson. I wouldn't see myself at their level. Would some people be saying: 'Who's she, or why is she putting herself forward like that?'

The programme is only thirty minutes long, but of course the interviewing went on for a lot longer. Gay Byrne was extremely nice and extremely gentle, a true professional. I was a nervous wreck the night it was broadcast because I didn't see it in advance. I remember getting a big mug of tea and sitting down to watch it with great trepidation because I didn't know how I'd come across to the audience. But I was so relieved when I saw it, and then we had to watch it all over again for the bits we missed the first time around. It's hard to take it all in, the first time, because there's so much going on. Something is said, and it sets you off thinking, and then you've missed the next bit.

I was on a roll with television then, and I went on *The Late Late Show* with Ryan Tubridy in September 2016. He was very nice to deal with. He'd put you at your ease very quickly and he was exactly like he is in front of the cameras. I suggested that he should have a go at ploughing and, fair play to him, he fulfilled his promise.

He was at the National Ploughing Championships in Tullamore anyway, because he was presenting his RTÉ Radio 1 morning show from there. As soon as he got that out of the way, he was raring to go. Great commotion surrounded him when he got on the overalls and headed to the competition area. He was in an open-top jeep and everyone was cheering as he went by. My son, DJ, gave him his tractor and plough, along with a few instructions. He played a blinder for someone who wouldn't know one end of the plough from the other. People couldn't believe that he had even got up on a tractor. And I think people thought more of him when they saw he was willing to do it. We raised €4,000 for charity from his ploughing debut. McHales, the machinery people, said they'd give €2,000 to two charities picked by Ryan Tubridy and ourselves if he went ahead with it. So, Our Lady's Hospital for Sick Children in Crumlin and the Irish Cancer Society got the benefit of Ryan Tubridy's first ploughing experience.

I enjoyed the small screen, but I must say I enjoyed the big screen even more. You might be surprised to hear that I had a brief brush with Hollywood. Very brief, mind you. It culminated in me walking down Fifth Avenue in the St Patrick's Day parade last year. It also led to me singing 'Sweet Sixteen' with Finbar Furey – a very fond memory indeed. Finbar was at the National Ploughing Championships in Ratheniska in 2015 for the launch of a trailer for the movie in which I had a very, very small part. Blink and you'll miss me. Of course, when people saw Finbar on the stage, they started calling on him to sing something, so he started on 'Sweet Sixteen'. I went over and joined him. It was a spontaneous thing, but it was really lovely.

But back to the movie. It all started, as most things do, with a phone call from someone I'd never met. Louth man Willie Martin called and told me he was making a film about Wild Goose Lodge and asked if I would be an extra in it. He said I was a woman of the land and he had admired what I'd done with the Ploughing.

Now, I'd never heard of either Wild Goose Lodge or Willie Martin. But after being in the film, I can give you chapter and verse. Wild Goose Lodge, in Reaghstown, Ardee, County Louth, was owned by the Lynch family and was burned down on the night of 29 October 1816. Some eight people died in the fire, including an infant. The fire was seen as vengeance for the hanging of three people, which had followed an investigation into a burglary in Wild Goose Lodge a few months earlier.

The fire led, in turn, to the controversial trial of eighteen local men – many of them innocent – and they were all hanged. It is a terribly sad story, when you think about it, and it has always been a big story in County Louth, but Willie wanted to tell it to a greater audience.

So off I went to Reaghstown to be an extra in a barn dance. Anna Marie was involved too, so my sister Eileen came along to mind Anna Marie's son, Saran. It was being filmed in Knockabbey Castle and everyone was dressed in the traditional clothes of the day. I had a white bonnet, long black clothes and laced-up boots. Eileen and Saran were kitted out too in the clothing of the day, even though they were only supposed to be there as spectators.

Dave Duffy, who plays Leo Dowling in *Fair City*, played the central role of the priest, Father McCann, and John Connors, who has been making all those documentaries about Travellers, was in it too, as well as the comedian and actor Joe Rooney, who will be known by fans of *Father Ted* and *Killinaskully*. Finbar Furey was singing and acting in it.

It was great sport and when I went home, I thought that was the end of it. But I didn't know Willie. He kept saying he was going to bring the film to America, and I just thought it was fanciful talk. Anyway, he rang me in February 2016 and said we were off to the premiere of the film in New York. I didn't believe him at first, but he had secured a screening of the film on Broadway. He is a very determined man. So, we went along to America and in we got to the stretch limos for the premiere. I had to pinch myself a few times to make sure I wasn't dreaming. It was wonderful.

The trip was timed to coincide with St Patrick's Day. We

JJ Bergin, co-founder of the National Ploughing Championships and my first boss.

Denis Allen, co-founder of the National Ploughing Championships.

Seán O'Farrell, who I succeeded as managing director of the National Ploughing Association.

This was taken by an Irish Press photographer in the early 1960s, when I was getting some media coverage for my role in the NPA.

The glamour of it all. Here we are, boarding the plane for the World Ploughing Contest in Rome, in 1960. NPA managing director Seán O'Farrell is third in from the right in the front row. (Barry Mason Photography)

Me (*front left*) with members of the World Ploughing Organisation at the World Ploughing Contest in Killarney in 1954. JJ Bergin, NPA founder, is directly behind me on the left, beside JD Thomas, president of the World Ploughing Organisation. May Bergin, JJ's daughter, is next to me.

The Republic of Ireland's first ever World Ploughing Champion – Charlie Keegan, 1964.

At the home place in Clonpierce in the late 1950s, after I had become the NPA's company secretary. The loft behind was my office. (Irish Times)

The NPA Central Council members on the occasion of our Golden Jubilee celebration, in 1984.

A typical working day – glued to the phone in the office at my home in Fallaghmore, in 2004. (Alf Harvey)

Martin Kehoe (*second from left*), the only three-time Republic of Ireland world champion, at the 1995 World Championships in Kenya. Martin's late wife, Karen, is to his left. Also pictured are Martin's team-mates: (*left to right*) Charlie Bateman, Jimmy Grainger, Ben Buttimer (RIP). I'm on the right.

President Paddy Hillery at the National Ploughing Championships in Knocktopher, County Kilkenny, in 1978. Fianna Fáil TD and Minister for Agriculture Jim Gibbons is on the far left, and Seamus Whelan, father of political analyst Noel Whelan, is beside him.

At the Farm Tractor and Machinery Trade Association Show in Punchestown, County Kildare, with President Mary Robinson, in the early 1990s.

Taoiseach Bertie Ahern presented me with my Rehab Person of the Year award in the Burlington Hotel in 1998. (Lensmen)

Putting the best foot forward with Taoiseach Brian Cowen and NPA Chairman Eddie Hally as we walked the competition plots at the Ploughing in Athy in 2010.
(Alf Harvey)

President Mary McAleese cutting the ribbon at the Ploughing in Ballinabrackey, County Meath, in 2003, with NPA chairman Michael Mahon, Martin McAleese, NPA president Gerry Baker and myself. (Alf Harvey)

At Áras an Uachtaráin with President Mary McAleese and Anna Marie for a reception to mark the eightieth anniversary of the National Ploughing Championships, in 2011. (Alf Harvey)

The NPA council, and friends, celebrating eighty years of the National Ploughing Championships, at Áras an Uachtaráin. (Alf Harvey)

This is what it's all about – ploughing competitors hard at work at the National Ploughing Championships in Ratheniska, County Laois, in 2015. (Alf Harvey)

A small selection of the crowd at the National Ploughing Championships in Screggan, County Offaly, in 2016. (Alf Harvey)

Enjoying a quiet moment with Sabina Higgins, NPA chairman PJ Lynam and President Michael D Higgins at the National Ploughing Championships in Screggan, in 2016. (Alf Harvey)

In hallowed company, with the former Papal Nuncio to Ireland, Archbishop Charles Browne (*left*), the Bishop of Kildare and Leighlin, Dr Denis Nulty, and then NPA chairman PJ Lynam at the 2015 National Ploughing Championships in Ratheniska, County Laois. I thought it was a tremendous honour when Archbishop Browne came to visit us. (Alf Harvey)

Running a meeting of World scrutineers in York, in 2016. From left: Ismo Nakki, Finland; Ann O'Shea, Dep. General Secretary; Anna May, Irish Board Member (chair of Scrutineers); Richard Pedersen, USA; Bekim Hoxha, Kosovo.

Mary Brennan of the NPA team. (Alf Harvey)

Mórag Devins of the NPA team. (Alf Harvey)

The NPA team pictured with Ryan Tubridy at the Ploughing in 2016. (Alf Harvey)

Calculating the depth of Ryan Tubridy's ploughing when he tried his hand at the sport in Screggan, in 2016. He measured up well! (Jeff Harvey)

The Irish competitors always punch above their weight at the World Ploughing Contest. Eamonn Tracey came second in the 2016 conventional contest in York. His father, John Tracey, was a six-time winner of the rose bowl Eamonn is holding. John Whelan came third in the reversible competition in York. They are flanked by our judge, Brian Mahon, and coach, Declan Buttle.

Some of our site team (*left to right*): President James Sutton, Billy Gray, Tom Kelly, Louise Brennan, Sean Byrne, Pat Walsh, John Burgess, John Moran. (Alf Harvey)

Helping then Minister for Health Leo Varadkar to find his way . . . around the Ploughing at Screggan, in 2016. (Alf Harvey)

Another highlight of my life – meeting Cardinal Timothy Dolan at St Patrick's Cathedral when we went to New York for the St Patrick's Day parade in 2016. (Willie Martin)

Anna Marie and I are truly blessed among men. This is the council of the National Ploughing Association after its AGM in Abbeyleix, County Laois, in April 2017. (Alf Harvey)

found ourselves walking in the New York parade under the Louth flag, with Finbar Furey, Dave Duffy, lots of people from the film, and the weather man Gerry Murphy, who acted as MC for the Broadway and Dublin premieres. Every so often someone in the crowd would recognize Dave and shout: 'Howaya, Leo Dowling!' It was a long walk, but it was such a privilege and an honour to be there.

The former US senator and Northern Ireland peace-broker George Mitchell was the Grand Marshall of the parade that year. Cardinal Timothy Dolan met us at St Patrick's Cathedral and he was so welcoming. You'd feel you knew him all your life. We were brought right up to the front of the cathedral for the Mass. We just felt elevated by the whole experience.

The Irish premiere was held in the Savoy cinema in Dublin and that was another great night. I was delighted to catch up with RTÉ's Mary Kennedy on the night because we've crossed paths many times at the Ploughing. Because he was the MC for the event, Gerry Murphy brought all the weather girls to the premiere. He was a fantastic host and wonderful company. His grandmother was from County Louth, so he was very interested in the story of Wild Goose Lodge. The Dublin premiere was exactly what you would expect a film premiere to be — red carpet, flashing cameras, and everyone dressed up to the nines. Hundreds of people came from County Louth to see the movie and there was a great sense of camaraderie with everyone. The film was very well received on the night and it went around the country to a number of cinemas afterwards. We all went off to the Gresham Hotel for snacks after the premiere. I felt like a million dollars because when I was young, the Gresham was the epitome of glamour and poshness. If you had as much as a cup of tea in the Gresham, you would be bragging about it for weeks.

Willie Martin is a great character. He had worked for years to bring the film into being and you'd have to admire that determination.

Sr Consilio, who runs the Cuan Mhuire addiction treatment service, is another person I have tremendous time for. She is a Sister of Mercy and she was very close to John's sister, Sr Dominic. The two of them worked in St Vincent's Hospital in Athy together. They were so close that when Sr Consilio's father came to visit her from Kerry, he would stay in our house. Sr Consilio didn't have it easy when she started working with alcoholics and drug addicts, but Sr Dominic stood by her all the way.

It all started when Sr Consilio took some alcoholics from the streets and brought them into the convent in Athy, out of compassion. Now, not everyone agreed with her, and she was criticized quite a bit for her actions, but she stuck with it because she felt it was the right thing to do. The numbers of people arriving into the convent grew, and then she started Cuan Mhuire, a small drop-in centre about a mile outside Athy. Today it has grown to house a garden centre, restaurant and shop and she has a fantastic garden growing vegetables, which also provides employment. She has several Cuan Mhuire centres around the country now, in places like Ardee and Limerick. Sr Consilio has made such a difference to families throughout Ireland by helping people to give up their addictions. I really admire her faith in God and her devotion to Our Lady. She might be in serious need of money for something and instead of panicking, she'd just say, 'Our Lady will provide', and Our Lady always did.

Not many people know that Sr Consilio is a fantastic cook. She made my wedding cake. It was a four-tiered affair and she put a mirror under each tier so from a distance it looked like the tiers were resting on water. I couldn't find adequate words to say what I think of Sr Consilio. I couldn't praise her enough. I think she is Ireland's Mother Teresa, although I know it would embarrass her to hear me saying that. She has devoted her life to helping others and it has made a massive difference to so many families.

I love the way you meet all sorts of people in this job that you wouldn't normally bump into. But they're not always that nice. I remember meeting this judge at an event. He was standing on his own and I felt sorry for him, so I walked over to talk to him and he just turned his back on me. I thought to myself: 'Well, it's no wonder no one is talking to you, sir, if that's the attitude you have.' You'll always come across people like that, who have a great opinion of themselves and offer their hand with the air of, 'Here's my hand, shake it yourself'. I have no time for that kind of nonsense.

Of course, it's great fun meeting people who are in the public eye and seeing what they are really like, but I'm more grateful to the Ploughing for bringing me into contact with people who really are the salt of the earth.

I'm thinking about people like Carrie Acheson, from Clonmel, who has been the voice of the Ploughing for nearly thirty years. She sits up in her perch at the event and does all the announcements in that lovely tone of voice. People love to listen to her. She must have reunited thousands of lost children with their parents over the years. And as she always says, we've never left a child behind yet. It's funny how lots of people actually think that I do the announcements. I have often been asked: 'Sure how do you do it all? We see you walking the trackway and the next minute we hear you on the tannoy.'

And then there's my brother Oliver's wife, Mary Brennan.

She worked with me at the Ploughing for thirty-three years and we've never once had a cross word. We went through thick and thin together, back in the days when there was only the two of us in the office and things might not have been too easy. Mary had that rare ability of being able to get what she wanted while leaving the other person satisfied and feeling that he had got his way. She could peel a guy like an onion and yet he would go away happy. Mary retired two years ago, but still people ask for her all the time. The ploughing people loved her and it's people like Mary who give the event its lovely warm atmosphere. She is now enjoying lots of precious time with her grandchildren and is as busy as ever.

And I'm not forgetting Séamus Whelan, who was our exhibition supervisor for thirty-six years. His son, Noel, will be well-known to many as *The Irish Times* political columnist and barrister. We always think of Noel as one of our own because he came to us for work experience as a young lad. I remember collecting him off the train. He was well able for anything we asked him to do and was always anxious to learn. I love to hear him on the radio now. He's a great analyst.

His father, Séamus, was a big Fianna Fáil man – he was a councillor in Wexford and he was assisted in his Ploughing task by a Fine Gael man, the councillor John Moran. They used to have great sport when they were planning where the Fianna Fáil and Fine Gael stands would go in the exhibition arena. Of course, it was just good-natured banter. They knew there would be little difference in locations for political parties, but they loved to rile each other about it. John is still heading up the team, which works from July through to the end of September, mapping the ploughing on the ground.

There are so many people in the NPA down through the years who have guided and shaped me into the person I am today. I have always felt the strength of the Association at my back in everything I do in the name of the Ploughing. Whatever the crisis of the moment is, there is always someone to call for an opinion – from legal advice to financial advice to personality clashes. We really have a power house of expertise in our midst and it makes my job easier when I take advice before reaching a conclusion. Now, don't be under any illusion that I do what I'm told – far from it – but it's great when you have access to a range of opinions. You're more likely to reach the right conclusion that way.

Some ploughing families are like family to me because we go back so far. And I mean far. As I was writing this book, I went to Killarney for a meeting about retrospectively erecting that Cairn of Peace to mark the 1954 World Contest. I was told there were a few people who would like to meet me. I couldn't believe my eyes when I saw the gathering – it was a group of men and women who were part of the 1954 event. I had not met many of them since then. That event may have been sixty-three years ago, but I can tell you the few minutes I had with that group brought me right back to my first World Contest.

Of course, we have our differences of opinions at NPA meetings, so don't get the impression that it's always rosy, but at the end of the day I'm very fond of them indeed and they'd go through thick and thin for the Association. That spirit is summed up by Carlow man John Tracey. He was due to represent Ireland at the World Ploughing Contest in Australia in 1997, but when we were travelling to Dublin Airport, I got a phone call from our travel agent to say that John's mother had died. We had to tell him when we met him at the airport. Everyone was so sad for him but, despite the bereavement, John came out to Australia after the funeral and ably represented Ireland. I think that really highlights the dedication of ploughing people to the competition.

There are a few people I'd still love to meet. The Pope is coming to Ireland next year and wouldn't I love to have him at the Ploughing? Unfortunately, the dates of his visit don't suit us. I did check though, just in case! Wouldn't it be wonderful? The Pope-mobile would look great going down our steel trackway. I suppose you could say we had Pope Francis in spirit at the Ploughing last year when the Jesuit pavilion had a life-size cut-out of him. I heard that people were queuing to get selfies with 'the Pope'. We had sixteen religious stands at the most recent Ploughing in 2016, representing various religions, orders and religious groups. Some have prayer rooms and quiet spaces for reflection, which are very useful when you need to take a breather after battling through a crowd of 80,000 or 90,000 people.

When we had our AGM in April, I told our directors that while we couldn't get the Pope, I was considering inviting US President Donald Trump to the Ploughing if the dates coincided with his expected visit. Of course I was joking, but they nearly turned the table over with the indignation! Could you imagine bringing himself and Melania down to the ploughing plots? Lord above, we'd have to have a terrible amount of security. Although I'm sure they'd pick up a few bits from the trade stands, and maybe a hurley for their young lad, Barron. No one goes home empty-handed from the Ploughing.

13. Becoming a widow

I don't mind saying that I was glad to see the end of 2007 because it wasn't a good year. John, my husband of more than forty years, died suddenly on 17 January. He was eighty-five years old, but he was a very active and lively man and you wouldn't think he was that age. His death was such a shock to us all.

John was a late developer of diabetes and he was sent to St James's Hospital in Dublin for a procedure. Myself, Anna Marie and her now husband, Declan Buttle, had travelled up with him. He had a clot on his knee and had been given something to move the clot. He had to remain on his back for twenty-four hours, so Anna Marie stayed in his room overnight to make sure he was comfortable. I arrived at the hospital early the next morning, while he was being prepared for theatre. Then DJ and his wife, Fiona, arrived in and they told us to go away for a cup of tea and they'd stay with John because he would need us later.

We had only got as far as the car park when Fiona called and asked us to come back to talk to the doctor about the procedure John was going to have done. When we got back inside, the consultant, Mr Dermot Moore, met us and told us that John had passed away with DJ and Fiona by his side. Mr Moore was so kind to us, and he had been very attentive to John. He told me John had never reached the theatre. The blood clot went to his heart. It really was very sudden. I remember the staff had brought him a salad tea the evening before and John said: 'Take that stuff away and bring me a fry.' And they gave him a fry. And he enjoyed it. I'm glad he did. Would you believe it, he was humming 'The Stone Outside Dan Murphy's Door' that evening? He loved singing that song because there was a bit of a go to it. And also anything by John McCormack. He was a great favourite.

John was very good-humoured and in such good form that night. It was a dreadful shock. I was broken-hearted leaving the hospital. Fiona drove me home and I think it hit me worse when I got back to the house. My immediate family and the girls in the office were already waiting for us in the house and they had started making preparations. It's really at times like this when you know the importance of family and good friends. All I wanted to do was to bring John home to the house and have him surrounded by the people who loved him.

Seeing the clothes hanging in the wardrobe and his bits and things around the house was very difficult. I kept them for ages, and I suppose it doesn't help to do that. But I still have some of his things, like ties, handkerchiefs and shirts, and I'd think that maybe DJ would use them some day. I couldn't get rid of his pipe and his glasses. They are still sitting in their spot on the windowsill, behind the kitchen table, and they'll never be moved.

There was such an empty space when John left us. It was about one month after he died that the terrible loneliness hit me. I'd catch a glance of his empty chair and it would nearly knock me sideways. Or I'd think of something he would have said or done and then I'd remember.

John was a great sounding-board. Often I'd come home a bit irritated by some remark someone had passed. I'd go up the yard where John was working and tell him what happened. And I'd say I felt like writing a letter to them and he'd say: 'Look it, Mam, leave it 'til the morning and see how you feel then.' (After the children were born, we always called each other Mam and Dad.) It was always good advice to take. The next morning would come and the thing that was said wouldn't seem as bad at all and I'd have a completely different view of it. The night would have taken the sting out of it.

We were very different in the way we would react to things. The incident with the pet lamb, Daisy, comes to mind. I had been cultivating beautiful pink roses just outside the window and I loved looking out at them. Daisy was prone to wandering around the garden, as pet lambs do. But didn't I look out one day to see her taking the heads off my beautiful roses? Well, I lost my head entirely and stormed off to John. 'Either she's going or I'm going,' I said. 'Go off with yourself so,' he said, as calm as anything. He had a twinkle in his eye, though. He didn't get a bit excited, while I was ready to hit the roof.

That same lamb saved us from a serious burglary after demolishing my poor roses. Anytime Daisy heard noise in the garage, she would start belting the door with her head because her bag of feed was in there. Anyway, this night we heard noise and came out to find Daisy in the garage, where the Wolseley car was kept at night. The car doors were open and the glove compartment ransacked. A bicycle was gone. We concluded that Daisy had heard the commotion and started banging on the door for feed. The thieves thought it was someone and fled with the bicycle. We found it in a ditch up the road. Two cars were stolen in the locality that night. So I couldn't complain about Daisy after that, could I?

John loved the Ploughing, but he always kept in the background. He used to think that I worked too hard. If he saw me working late at night, he'd say: 'You'll kill yourself at that work. Can you not finish at a reasonable time?' He was a very quiet and patient man. I suppose he wouldn't have stuck with me for so long if he didn't have patience. But he was also genuinely supportive. If I was going anywhere for a night meeting, I always knew that John was there with the children and there was nothing to worry about. I remember Anna Marie got him a little plaque once that said, 'I love you Dad, because you are always there', and that truly summed him up.

He was a real home-bird. I suggested going to Lourdes once and he said: 'Haven't I Lourdes in my own home if I kneel down and say my prayers at night-time?'

As well as being there for Anna Marie and DJ, John was always there for me too, so the time after he died was very sad. The one thing that kept me going was the support from everyone. And were it not for the Ploughing, I don't know how I would have coped. The work kept me focused. I still had to be up every morning, doing something for the Ploughing, and it took my mind off things. And then people were always calling to the house because of the Ploughing. Because the office is attached to the house, there's always plenty of activity. You know how houses can get so lonely when someone dies and people gradually stop calling.

Anna Marie was living with me and we had DJ and Fiona down the road with their first baby, so I was lucky to have everyone around. And it was great that John got to see his first grandchild, Seán Óg. He was nine months old when John died.

I remember someone saying to me, just after the funeral, that I should think of all the things that John used to say and jot them down. Then I should take them out when I felt able, and look back on them. I did that, and it helped. You forget things about people when they are no longer with you, so it was a real comfort to look back on what I'd written down.

Matt Dempsey, former editor of the *Farmers' Journal*, wrote a lovely piece about John's funeral afterwards, mentioning how the homily highlighted John's love of farming and football. 'After the Requiem Mass, the coffin was carried shoulder-high half a mile to the adjoining cemetery on the hill, overlooking Co Laois and south Kildare and just a few miles from John's farm and home at Altona,' he wrote.

I was so proud of the send-off we gave John. All our nephews and godsons wore black suits and ties as they carried John's coffin. The Charlie Landsborough song 'What Colour is the Wind?' was sung at his graveside, and I don't ever remember shaking as many hands as I did over those few days. I can tell you, I was very humbled by the number of people who came to support myself and my family. I hope the great Irish tradition of celebrating our dead never changes.

The members of the National Ploughing Association were so good to me during and after John's funeral. Our AGM came up a few months later and I found it very emotional when I stood up to thank all the directors there who were so kind to me.

You know, quite a few years have passed and I still have not got a memory card made for John. I can't say why. I think perhaps it's because I could never bring myself to see him in that way.

There are so many milestones when a loved one dies: the month's mind, the anniversary, Christmas, Father's Day, their birthday. And, of course, that year's Ploughing Championships was a big milestone for me. It was in Tullamore, and I found it very hard leaving the house for the Ploughing and knowing that John was not there and he would not be following behind me. He always came to the three days of the Ploughing. My sister Eileen used to help him to pack and she would always take out a good suit for him, for the presentation of the awards. Without fail, John would say: 'What do I want a suit for? What would a man like me be doing, going to a thing like that?' Then Eileen would let on that she was taking the suit out of the case, and he'd say: 'Sure, throw it in anyway, just in case.' And of course he'd go to the awards. He always did. He loved the Ploughing.

John and Eileen's husband, Tommy, would always park themselves on two chairs outside the NPA headquarters at the Ploughing and watch the world go by around them. John would be quite content, puffing away on the pipe, and he'd be able to tell us about things that we didn't even know were going on. They had no interest in meeting the VIPs. 'I don't want to be talking to those lads,' he'd say if he heard about important people coming.

All these memories were going through my head when I left for the Ploughing without John for the first time. But I forced myself to keep putting one foot in front of the other because I knew that's what John would have wanted me to do. And you get through it and it gets easier.

We pass the graveyard where John is buried on the way into Ballylinan, and there are just two graves separating John's grave and my parents' grave. DJ would often say, 'Howya Grandad', as he's going by with the children on their way to school. The grandchildren think they remember him because we talk about him so much. I truly feel John has not left me as I still feel his presence around. I pray to him often, asking him to look out for me and the children and grandchildren. I really do feel a great companionship between us even though he's gone more than ten years now.

As every widowed person will tell you, becoming a widow changes things a lot. The loss was most noticeable when I would return home from a meeting or an event and John was not there to talk to. He was such a great adviser, even though he appeared to be a very relaxed, laid-back kind of person. He had a real calming effect. Things would happen and I'd say in my mind, 'Wait until John hears that', and then I'd remember he wasn't there anymore.

John would have always come to functions with me and was great company, so that was another thing I greatly missed. It's not the same, going on your own. Even now, when I go to weddings, I really miss him being there. He was a great dancer. He'd glide around the floor with a lovely elegance.

I know I'm luckier than many people who have been widowed because Anna Marie, her husband Declan and son Saran live with me and DJ is here every day farming. It must be very hard for widowed people who don't have any family around them. I can't imagine how lonely it must get.

I know John was inwardly very proud of what I'd achieved with the Ploughing, but he'd never tell you that. Not at all! Someone would come in and say to him, 'God, you must be very proud of Anna May', and he'd say, 'Now you wouldn't want to tell her that or she'd get a swelled head.' That was the attitude there. One day he said to me, 'You know, Mam, I think you could be dead a month before I'd miss you because you're always away.' He was very witty. But he used to say to me, 'You'll kill yourself at that Ploughing. Is there no clocking-off time for you?' I'd have to work late into the night to get things done before the phone would start going in the morning. Saturday and Sunday meant the same to me as any other day. You wouldn't believe some of the phone calls I'd get at night-time. People would ring asking, 'When did so and so win the Ploughing?' or, 'Where was the Plough-ing in such a year?' and I'd know straight away they were at a quiz somewhere. You'd hear the noise in the background.

Mind you, John never suggested that I should retire. He knew he couldn't live with me if I did.

That year, 2007, was also the year when I lost my brother JJ. That came just three months after John's passing. I was still trying to cope with the sudden death of John when JJ went. It hit all the family hard, but it really floored me. As I mentioned earlier, JJ was a great support in my Ploughing work and he was a great confidant. I always respected his opinion and looked up to him, like many people look up to their big brothers. JJ had been sick for a while and was in the early stages of Alzheimer's disease, but his death was still a huge blow.

On the evening of his removal, I remember Anna Marie coming into the house and asking why I wasn't at the family home place, because that was where JJ was being removed from. I was sitting on the second last step of the stairs and I told her I just couldn't face it. It was too much heartbreak.

I had to summon a deeper strength from somewhere. I forced myself to get up and I went with her to see my oldest brother leaving our family home for the last time. It was the hardest few months of my life, to lose John and JJ so close together. They were great pals, too.

When you're young, you think that grief doesn't affect older people, that they should be used to it, but that's the folly of youth. Losing a sibling is tough at any age and noth-ing prepares you for the upset. You have such a shared history together and no one knows you in the way that your brothers and sisters know you. The only advice I can give to anyone in these circumstances is to lean on your family and dear friends. They will help you to get through it. Every death of a family member or friend takes a little bit out of you, but time does help. It's sad to look back on old photographs and see how many people are gone, but you can't dwell on that because it doesn't do you any good.

Paddy was my first sibling to die. He had chest problems for some months, probably farmers' lung disease, and he passed away in 1999. The first death in the family is a big wrench for everyone and we really felt it. My brother Gerald died suddenly in 2011. He had been in good form, watching sport that night on the television, but he got up during the night and collapsed. He had a stroke and he died the next morning.

I have to say, my brothers all married wonderful women. We have great sisters-in-law and they're always there whenever they're needed.

I had another setback in April 2016 when I was involved in a head-on collision. It was a beautiful, sunny evening and I was heading off to Portlaoise to present school certificates. I was going into Stradbally when I saw this jeep heading across the road towards me. I had nowhere to go because there was a stone wall on the other side, so I took a tight grip of the steering wheel and said to myself, 'I've had it, now. That's the end of me.' There was no way I could escape, so I slowed down as much as I could and braced myself for it.

The collision shattered the driver's window and set off the airbag. When the gas came out of the airbag I thought it was smoke, so my first thought was to get out of the car as quickly as possible. I got the seat belt off and I pulled myself out through the broken window, one shoe on and one shoe off. The front wheel was gone off my car so I didn't have a big drop on the other side.

I'll never forget these two kind women who came out of their houses and gave me a chair and wrapped my legs and feet in a blanket. The poor chap in the car behind me was hit as well.

The ambulance people were so nice to me and I remember one of them saying, 'Wouldn't it be worse if it happened when you were on your way to the Ploughing Championships?' That really put me at my ease.

I was extremely lucky, not a bone broken and, thank God, no one else was seriously injured. I'd had a hip replacement two years earlier so we were worried that the hip would be affected, but it was fine. The worst thing was the bruising. The seat belt left massive bruising and it lasted for six or seven months. If I hadn't been wearing my seat belt, I would definitely have gone through the windscreen.

Someone was looking out for me that day. I always say it was Our Lady who saved me because I have great devotion to her. Any other day, I could have had Eileen or my sister-in-law Anna with me and if I had, maybe the outcome would have been a lot worse. It took me several months to get the courage to get back behind the wheel again. People were so thoughtful after the accident and so many people surprised me with their acts of kindness. I still remember seeing this very senior Garda officer coming in to visit, and him swinging a bag with a nice carrot cake from the Stradbally Fair in it.

Our two children got into the Ploughing – sure they had no choice, I suppose, because they were surrounded by it. But I never encouraged either of them to get involved. When I say this, Anna Marie always says: 'What did you expect? The Ploughing was talked about morning, noon and night in our house.' She's not exaggerating. They never heard anything else but Ploughing in our house. I don't know how John put up with us.

DJ was interested in ploughing from a very young age. When he was fifteen, he won £1,000 from Prize Bonds he had been given for his Confirmation. There was only one thing he was going to do with that money – buy a plough. He was at a great age to get interested in ploughing. So off we went to Larkins of Baldonnel, where he bought this lovely Kverneland competition plough. It would cost about €7,000 to buy a plough like that today. He still uses that same plough, although he has upgraded it plenty since then. He loves ploughing and I've heard him say he gets so immersed in the work that when he gets off the tractor, he doesn't know what day of the week it is.

DJ is very involved in the ploughing organization in Laois. He was chairman of the Laois Ploughing Association when we held the National Ploughing Championships for three years running in Ratheniska, from 2013 onwards. He's great for getting younger people involved in ploughing. Back in the 1990s, his local ploughing branch introduced the novice match for young people aged over sixteen who are interested in ploughing. Now every county has a novice match and some of those youngsters have come up the ranks and won a few All-Irelands since.

Anna Marie also started ploughing when she was a teenager and she first entered in the farmerette class in the National Ploughing Championships in 1989, in Carlow. She was joint sixth on that occasion. She gradually crept up the ranks and was placed on many occasions. Then, in Kilkenny in 2008, she was thrilled to join my sister Eileen in the history books when she won Queen of the Plough. It would have been a great day for John to have shared. He was very proud of the two of them ploughing, but he could never have been accused of being boastful about it. No matter how well DJ ploughed, John would just say, 'Yes, you did a fair job', but we all saw how delighted he'd get when either of them won.

Anna Marie always had a great interest in the office and she understood the workings of the Ploughing very well. So, when the NPA was looking for a World Ploughing Contest coordinator in 1993, it came as no surprise that she was interested. That role was for the 1996 World Ploughing Contest, which was due to be held in Oak Park, County Carlow.

Anna Marie had been working as a professional youth worker with Foróige, which she loved, but I suppose she saw that life was presenting its next challenge and she went for it. I was anxious to stay as independent as possible so she was interviewed by our national executive committee. She got the job, and her first task was to bring in enough sponsorship for the world event to cover her wages for the duration of the role. That was a major piece of work but she achieved it, no bother.

After that world event, Anna Marie did as many young people do and headed off on a world tour, ending up in Australia. When it came to travel, Anna Marie didn't take after her father at all. She was mad for travelling. I well remember the trauma when she went to waitress in the Isle of Man for a summer during college. She had a pet gosling called Douglas at the time and the only instruction she gave us on leaving was that Douglas had to be minded. The gosling would follow you all over the yard, even into the bed if you let him. I did wonder what would happen when he grew up. Would he be following us around the shops in Ballylinan and Athy like a dog? Anyway, Anna Marie wasn't long gone when the cat grabbed

Douglas by the neck. We tried a bit of emergency medicine on him and bandaged him up, but I'm afraid Douglas was a goner. We all thought it was for the best if Anna Marie didn't hear about the gosling's passing until she got home.

A while after, I sent DJ over to the Isle of Man to make sure she was doing okay. I even loaded him up with home-baking and an electric kettle for her. He brought a girlfriend with him and their express instructions were to avoid all talk of Douglas. I'd say they weren't five minutes in the place when Anna Marie asked if there was any news.

'Nothing,' said DJ's companion, 'except the dead gosling.'

The Isle of Man wasn't too bad, but it was a different story when she decided to go to Australia. She had a going-away party the night before she left, and for us it was more like a wake than a party. I remember John saying, 'Sure what notion is on her to go wandering?' He had come from an era where young people only left when they had no choice, whereas Anna Marie was choosing to take off to explore the world. He was very sad about that and I suppose he was worried that something might happen to her. And like all parents whose children go to Australia, he was wondering in the back of his mind if she'd come home after the travelling or decide to settle over there.

But happily for us, fate intervened. When she was off travelling, the NPA executive told me they wanted to offer her a full-time position dealing with public relations and coordinating competitors. Up to then, PR was always the responsibility of Larry Sheedy and he was fantastic, but the growth of the event meant that we needed to have someone in-house. Larry was a wonderful mentor to Anna Marie and is still a great friend to the NPA.

So Anna Marie has worked permanently with the NPA

since 1997, and is now one of three assistant managing directors, with responsibility for public relations and event
management.

In 2013 – with some considerable pressure from me, I have
to admit – Anna Marie applied for the position of general secretary of the World Ploughing Organisation. There had never
been a woman in that role and she was humming and hawing
about going for it. She honestly didn't know if she had a chance
at all, but I could see that she had as much experience as anyone else, so I egged her on. She faced an international interview
panel from five countries, in Amsterdam. It was a proud day
for the family when she was selected in April 2013 as the first
Irish, and the first female, general secretary in the sixty-year
history of the organization.

This was a wonderful personal achievement for Anna
Marie, of course, and I suppose it was her chance to prove
herself, independent of me. It was also a tremendous boost
to the NPA as we had not held any executive position with
the World Ploughing Organisation since JJ Bergin was
vice-president in 1954. She never looked back, once she got
the job. Now it is very demanding and she has to travel quite
a bit, which is hard with a small child. On top of that, she's
probably busier than ever in her job in the National Ploughing, but she is well able to handle it all.

It's interesting to see how Anna Marie and Declan manage childcare, with both of them working, and Declan
commuting to Wexford. I had great back-up with John and
my sisters when my children were young, so I was happy to
be able to give that kind of support to Anna Marie and
Declan when their baby Saran was born. I really feel for
young women who don't have their parents nearby when
they have their first child. There's nothing more reassuring

than having your mother nearby when you are dealing with a new baby for the first time and you don't know what you are doing.

Anna Marie has a fantastic support network as Eileen's daughter, Elizabeth, helps with Saran. And if both Anna Marie and myself are away, he often takes off to his auntie Aoife Redmond in Ballymurn. Sure, where would we be without family?

Like our own Ploughing, the World Ploughing has gone from strength to strength and today thirty-three countries are involved in the organization. I am now the longest-serving member and the only woman on the board. I'm chairwoman of the scrutineers' committee, which keeps me busy when we attend World Ploughing Contests. There are five scrutineers at the World Ploughing Contest and we spend our time taking the judges' sheets, inputting the results and double-checking them.

I was proposed as chairwoman of the WPO back in 2000. That job entails conducting board meetings, making decisions and passing them on to the affiliated countries. It was daunting, but I decided to let my name go forward because I looked around the room and knew I had as much experience as anyone else there. Things were looking very good for me after the first count, but for some reason they decided against eliminating the person with the lowest number of votes. I could see that we would arrive at a stalemate if people with the lowest number of votes were not being eliminated. A deadlock was inevitable, so I decided to withdraw to move things forward. Ken Chappell from England was eventually elected. I was half-sorry and half-relieved.

In this chapter I have talked about the various low points I've had in the last decade, but I've had more than my share of very happy times as a mother and grandmother too. See-ing Anna Marie and DJ finding their partners, getting married and having children made me very content. And would you believe it, both of them met their spouses because of the Ploughing. I think I should get a bit of credit for that.

DJ's wife, Fiona O'Donnell, is from Dún Laoghaire in County Dublin, but ploughing still managed to bring them together. She was great friends with the wife of a ploughman and she came to a ploughing dinner dance with them. That's where she first clapped eyes on the brave DJ.

Anna Marie met her husband, Declan Buttle, at the World Ploughing in Denmark in 2001, but it took two years before they went out on a date. It might not surprise you to hear that he's an all-Ireland ploughing champion.

But do you think I was told about the budding romance? Not a chance. And I knew his father very well. They kept it secret from everyone for a very long time. I suppose I could see Anna Marie's point. She was involved with the NPA and was going out with a ploughing competitor. What would people be saying? There would be a lot of comment, so they wanted to make sure it was going somewhere before telling people.

I had to laugh when she told us about the first time she went to meet Declan's parents in Wexford. Anna Marie was nervous about the whole thing, but into the house she went and who did she see looking down on her from the wall? Yours truly. There was a photograph of me making a presen-tation to Declan when he won the All-Ireland ploughing. She must have thought she'd never get away from me.

14. Me and my faith

I think you'll have gathered by now that religion was, and still is, a huge part of my life. We were all good Mass-goers when I was growing up, and we still are. I wouldn't be a holy Mary by any stretch of the imagination, but I go to as many religious occasions as I can. I'm secretary of the Altar Society, and we take it in turns to do the flowers that adorn the church on Sundays. I always do the flowers for St Anne's Church in Ballylinan at Christmas. It's a lot of work and it can often take a few days, but it's an honour to do it.

I honestly don't know how I would survive if I didn't have God in my life. I feel He is with me everywhere I go. I'd often say, 'Dear God, help me get through this situation', or 'Please God, help me solve this problem.' I've never doubted the presence of God, ever.

The first thing I do when I wake up every morning is to say my few prayers. I have great devotion to Our Lady and I pray to her for a lot of favours. I feel that my prayers are always answered, in some way. I love my rosary beads and I often fall asleep saying my rosary, waking up with the beads around my neck. If I was driving home from a meeting at night-time, I would either be singing or saying a few prayers. If I was on my own, that is. I would be getting some strange looks from some of my travelling companions if I burst into a good ballad, or 'Queen of the May', which is my favourite hymn.

I'm a great one for praying to St Anthony when I lose something and he doesn't normally let me down, as long as I

keep the contributions topped up. Prayer is a great support
to me, particularly in the run-up to the Ploughing. Before
the Championships I always pray to St Thérèse of Lisieux,
the 'Little Flower', for good weather. But more importantly,
I always pray that there will be no major accident. JJ Bergin
once declared that St Thérèse was patron saint of the Ploughing, so I've kept that up.

I climbed Croagh Patrick twice, in 1960 and 1961, and I
followed that with four pilgrimages to Lough Derg in the
following years, back when the pilgrimage was a terribly
strict thing. Younger readers might not know what's entailed
in that pilgrimage. It's three days of fasting and prayer, walking barefoot as you do the Stations of the Cross, and a
twenty-four-hour vigil with no sleep. Nowadays you can do
an easier version of the pilgrimage – a day-long retreat where
food is served and you keep your shoes on.

I did those trips before I got married. We'd head off
together, a gang of four or five friends. You'd start fasting at
midnight the night before you'd go, so we'd be starving by
the time we'd get up to the island, off Donegal. I'll never
forget the smell of the fry coming from the presbytery in the
morning as the priests got their breakfast. And of course
they had to eat because they lived there for the whole summer, but we felt very hard done by, getting the whiff of the
sausages frying.

I remember Monsignor Flood going around giving us a
poke on the shoulder if we were lying back to get a bit of rest
during the day. 'Sit up straight,' he'd shout and you wouldn't
be long straightening up for him. At night-time you'd be
wishing that you'd get a good strong preacher at about
3.00 am, when the sleep was fairly hitting you. A priest bel-
lowing from the pulpit would keep you awake much better

than someone with a quiet, droning voice. The first time we went to Lough Derg, one of my friends came up with a plan to get a seat in the corner, at the back of the church, for the night vigil. She was going to roll in under it and sleep to her heart's content when no one was looking. That was before we met Monsignor Flood and realized what was involved.

You think you'd eat rings around you when you were finished the fast at midnight on the third day, but after a few mouthfuls of food all you'd want to do was sleep.

I thought it was going to be a holiday the first time I went to Lough Derg, but I soon revised my opinion. I went back, though, and that's because I felt a great sense of satisfaction once it was over. There was a great lightness about you afterwards, and you felt like you'd achieved something.

Lough Derg and Croagh Patrick were challenges, I suppose, in the same way that those endurance runs and cycles are now. When you're young, you're keen to test yourself and see what you're capable of. You'd hear of other people doing it and say to yourself: 'Why not? I could do that too.'

Religious events were at the heart of our lives growing up. We loved to see the mission coming to the parish, with its preachers, every few years. They were run by religious orders such as the Redemptorists. The missioners would set themselves up in the parish for a week to celebrate several daily Masses, hear confessions and give very strong sermons about morality. The stalls selling religious ornaments and trinkets were set up in the churchyard and we children could have spent hours looking at the rosary beads and statues for sale. And it was such an honour if you were asked to help out behind the stall if the stallholder was very busy.

There always seemed to be two lovely priests and one very serious one. And Lord, when the serious one gave the sermon on courtship and company-keeping he'd nearly be rattling the altar rails. We certainly wouldn't miss that sermon! Sure, he wouldn't be listened to nowadays, but back then there wouldn't be a sound in the church when he was preaching. And the church would have been packed with young people at that time. We'd even have new outfits for it.

We were very much afraid of the parish priest when I was growing up, but now I feel we are all on level terms and that's a healthier way to be. A lot of things have improved in the Church since I was a child. There was a Protestant church about two miles from us and, as children, we snuck into it with our cousins to see what it was like inside. We were petrified afterwards that our parents would find out because stepping inside a Protestant church was seen as a terrible thing. That was so wrong, when you think about it, and what did it achieve, at the end of the day? And I'm glad to see that mixed marriages are fully accepted now. I remember girls being put out of their homes for going out with a Protestant, many years ago. There was murder over mixed marriages.

I suppose the Bishop Eamonn Casey controversy marked the start of the change in the way the Catholic Church was viewed in Ireland. It affected how young people saw the Church, without a doubt. But I think you have to look at the balance sheet, and when you add up the good that Eamonn Casey did, like setting up the Third World charity Trócaire and helping poor and marginalized people, I think it outweighs the mistakes he made.

I never had any question about keeping the faith, not even during the various church scandals. I thought the church's PR department handled the abuse scandals very badly indeed. Its big failing was that it just didn't communicate with its people. While some priests did talk openly about it all, and congregations were all the better for it, other clergy-men just wanted to sweep it under the carpet and continue as normal. That made a lot of Catholics question what they believed in.

In fact, I went the other way when all the child abuse rev-elations came out. I got stronger in my faith. I remember saying to myself: 'Well, look, will I go to Mass? What's the point with all this going on?' But I weighed it up and found that it actually strengthened me. I felt I needed to go out and profess my faith and use it to try and get through this terrible time in the Church. I have often said a prayer or lit a candle for the unbearable experiences those poor women, children and their families had bestowed on them. I also pray for the individual members of the Church who are carrying out their ministries in the best way they can. They are all tarred with the one brush by association, and that can't be easy either. I often think that if you don't have your faith, what do you have really when all's said and done? You would like to think that you'll have a clergyman coming to you when you are in need. When people are dying in hospital, they need someone like that, even if it's just to hold their hand. The sense of touch is so important at that time, and you'd feel very sorry for people who don't have it, who might never have visitors.

It saddens me to hear of these moves to remove all traces of religion entirely from society. Taking saints' names off hospital wards and taking down crucifixes in public places is disappointing. What will we end up with if we take all refer-ences to religious belief out of our daily lives? Whether people like it or not, it's part of our history, but this move-ment to separate Church and State is trying to erase that history. Of course, just because Ireland is still predominantly a Catholic country does not mean that we don't welcome other religions into our midst, but does that mean we have to hide all signs of our religious beliefs? I strongly believe that I am what I was born into. That's my culture and tradition and that's what I want to pass on to my children.

For as long as I can remember, whenever we got a new car or machine, the first thing we'd do would be to bless it with holy water. I carry a little bottle of Lourdes water in my handbag and I still have the holy water font at my front door and back door, to dip my finger in when I leave the house for a journey.

I love to keep religious customs going. On All Souls' night, I leave a key in the door and I always leave a glass of water on the table for the poor souls in case they are thirsty. My mother did that and I want to continue her tradition. And we still bring home the blessed palm from Mass on Palm Sunday and put the branches in the outhouses to protect the animals.

Rogation days are another old custom that we still observe, although I'd say very few young Catholics have heard of them. Apparently, the custom has links with a French bishop called Mamertus, who is now a saint. His term of office in the fifth century was marked by every disaster you could think of – animals and people dying, earthquakes, fires, the lot. Mamertus decided to lead his people in three days of prayer and procession, asking God to protect the people and look after the crops. The disasters stopped and other bishops copied him. Years ago, congregations would march to the boundary of the parish, blessing every tree, stone and piece of land while reciting prayers.

The first rogation day is on 25 April and is a major rogation, and the three days before Ascension Thursday (the fortieth day of Easter, so it falls in May) are called the minor rogations. For us, it involves going around the fields and blessing the crops with holy water that had been blessed by the priest on Holy Saturday. These are all lovely traditions to hold on to and I hope my children and my children's children keep these things going.

We always start the National Ploughing Championships with an ecumenical service, where the various Church leaders bless the grounds, the competitors and the spectators. It's a special little ceremony and it brings everyone together.

One of the greatest highlights of my life was the audience with Pope John XXIII in 1960. Little did we know he'd be dead three years later. We were in Italy for the World Ploughing Contest in Tor Mancina, Monterotondo, which is to the north-east of Rome. The World Ploughing organizers asked the Pope to meet the country delegates, and it was a very exciting occasion for us.

Pope John XXIII was so lovely and pleasant. He was as broad as he was long, and I remember standing in the Vatican waiting for him to come in through this door and he surprised everyone by coming in a back door and walking up by us, his garments flapping around him.

He was known as the Good Pope for all the good deeds he had done – helping Jewish people flee the Holocaust, working with the poor and setting up the Second Vatican Council. I wasn't surprised when he was made a saint three years ago, along with Pope John Paul II. It was an amazing thing to meet him. I'll never forget the feeling I got. I genuinely felt that something came over me. It's hard to explain it. It was definitely a sense of being in the presence of holiness. I remember the same feeling when I went to see Pope John Paul II in Phoenix Park in Dublin in 1979.

I must have thought I was in heaven when Pope John XXIII came into the room. A hush descended on everyone as we waited on him to speak. I remember how he singled out the Irish visitors. He was one of fourteen children, from a farm a few kilometres from Bergamo in the north of Italy. He was once quoted as saying that there were three ways of facing ruin: gambling, women and farming, and that his father had chosen the most boring one. When he met us he spoke about the land and said how he had often stolen a few apples from orchards. He made it so lovely for everyone there.

I wore a black suit, black stockings and a black mantilla on that occasion because you had to show respect, but remember we were in Rome in September and it was really quite warm. As soon as we left the church, I remember flopping down on the kerb to take off some of the black, and not caring who saw me.

We had another first a few years ago when the then papal nuncio, Archbishop Charles Brown, came to the Ploughing in 2015. It was a real honour to have the Pope's ambassador to Ireland at the Ploughing. The archbishop is such a charming man, with a perpetual smile. He returned to the Ploughing, in Tullamore, in 2016 – he said he didn't get to see any ploughing the first time around because there was so much else to see. We were so disappointed when he was appointed to Albania in April 2017. I don't know much about the ploughing scene over there, but I do know the Catholics of Albania are lucky to have him.

Anna Marie laughs at me sometimes for this, but I would have a very strict code of ethics and morals. And it wouldn't just apply to my own family, as many Ploughing people know. We have to book rooms for many of our people at the Ploughing and, put it this way, if I'm paying for the bed, I

need to know who's going to be in it. And if they're not husband and wife, well, they can pay for their own room. I
remember this fellow was going to be acting as steward at
the Ploughing one year and he rang the office and would
only speak to me. His problem was that we had put him
down to share a room with some other steward.

'What's wrong with that?' I said, knowing right well, and
he told me he wanted to share with his girlfriend. 'Well that's
not happening,' I told him. Then he said he wouldn't be
steward if that was the case. 'That's your choice,' I said. 'You
know what to do. Turn around and go home if it doesn't suit
you.' But he didn't go home, I'm glad to say.

That's just the way I was brought up and while I have no
objection to what people do in their own private lives, when
it's my call, then it's my rules. I'm upfront about my morals,
and if people want to move around in the middle of the
night, that's up to them, but I'm not organizing it for them.

Then there was the time we had a lot of international visitors over for the Kverneland ploughing challenge in Carlow
in 1990. We started this contest in 1988 for people who had
won a world championship. The challenge was that they had
to use a standard match plough, new from the factory, without adding any of their own attachments to the plough.

Michael Patten, who is now one of Glanbia's senior directors, worked with Larry Sheedy's PR company at the time
and they were organizing the accommodation for the foreign visitors. I got a hold of the list and saw that we had very
few married couples, yet lots of men and women were sharing rooms. I said to myself: 'Well, they're not going to be
sharing rooms if I have anything to do with it.' And I
changed it all around, putting women sharing with women
and men with men. I remember Michael went to Anna Marie

in a panic, asking how on earth was he supposed to tell the Europeans they weren't sharing rooms because they weren't married? And she said: 'Well, you tell Anna May that.'

And that's how it was left, and I didn't apologize for it, but I felt bad for Michael having to tell them. Can you imagine doing it now? They'd tell you to mind your own business in six different languages.

Of course, I couldn't end this chapter on faith without telling you about the great Passion of Ballylinan. I've been in the Ballylinan Drama Group for years and we've done lots of plays. There's nothing like a good play with a strong story. Needless to say, we've done *The Field* by John B. Keane. We'd have to, as I probably know every field in Leinster at this stage. I played the widow who decides to sell the field and causes ructions.

I loved my time acting in plays like Peggy O'Reilly's *Purse*. I particularly enjoyed playing Lady Flimsy in *The Roadside* because I had to speak very fancy for that one and dress in a posh way. We always had a very strong cast of actors in Ballylinan and even went to our bishop's house for garments.

But we took things to a whole new level with the Passion Play. It all started in 1992. We had just done *The Field* and we were wondering what to do next. A few people from Ballylinan had gone to Oberammergau, in Germany, to see its famous Passion Play. Since 1634, the people of Oberammergau have been running the Passion Play every ten years. It tells the story of the death and resurrection of Jesus Christ. It came about after the villagers promised God they would produce a play depicting Jesus if He spared them war and bubonic plague.

There was no risk of bubonic plague or war hitting Ballylinan at that time, but it had given us an idea. Jokingly, we said, 'God, maybe we should put on the Passion Play.' Everyone laughed, but then a few people said, 'Yes, maybe we should.' So we started to think about it and when I got home I asked John what did he think. 'Are you mad, woman?' was all he said. We had no money. We didn't have a big enough stage, and we'd need a big cast of nearly 200 to get it going. The entire population of Ballylinan is about 2,000 and the Oberammergau production needs 2,000 people, between the cast, musicians and production people. Nonetheless I felt there was something stirring, so we called a meeting.

Five people turned up to the meeting. That was disappointing, but we didn't stop. We spread the word and got it announced from the altar that we were looking for people to get involved. Our then parish priest, Father O'Shea, said he'd do what he could to help. We held another meeting and more than twenty-five people turned up. And from there we really got going. A local school teacher, Sheila Graham, agreed to produce it as she had seen the real thing in Oberammergau. She whittled the length of it down from five hours to three hours and while it sounds long, the time flies because there is so much happening. The Oberammergau play starts at 2.30 pm and runs until 10.30 pm, with a three-hour intermission for dinner in the middle, but we thought that would be a bit much for Ballylinan.

The local Moran brothers made the stage, in Ballylinan Hall, which was a big undertaking, and carpenter Billy Ryan made all the crosses and the big table for the Last Supper. We went around to several churches and collected old vestments and we got plenty of remnants from shops. The costume crew would arrive with their sewing machines every night during rehearsals and get all the characters fitted. We even had rain coming down on the stage during the

Crucifixion scene, and thunder sound effects. Everyone shivered when the rain came. We had to get three-phase electricity into the hall to stage the lighting effects, which was a big deal at the time.

The greatest fun of all involved the men getting their false tan applied. Now that was hilarious. The men were all wearing sandals and togas, so you could see the ends of their legs and arms, and you couldn't have all that white, pasty skin on display. We had a make-up crew working flat out, topping up the fake tan every night with sponges. The crew was very professional and didn't tolerate any protestations – everyone who went onstage had to be made-up. Some of the men were very shy about getting the fake tan applied in the beginning, and others were afraid it was a bit girly. That diminished as time wore on, and in fact I think they got to enjoy it! People got to know each other and it brought a great sense of community to the village. I can't think of anything else that could have brought the community together so well, and that spirit is still very strong in Ballylinan. Young and old, people from all religions and none, and people from all walks of life, worked side by side. Not only was the whole parish involved, but many people from outside the parish also took part.

The first performance, in April 1992, was an outstanding success, and indeed every performance since. Every one of them has been sold out. Since 2005 we've been staging it every five years and we've already had people asking about the 2020 play. It has truly put Ballylinan on the map. Sheila Graham continued to produce it up until 2010, and then Bernie Dunne took over for 2015.

The men must grow their beards for months beforehand to look the part. I don't know what a visitor to Ballylinan would think if they dropped in to the village before the play,

with these huge bushy beards everywhere. I remember someone asking me if the men had gone stone mad in Ballylinan
because the beards were being commented on when they
went to cattle sales and such like. Of course, beards are the
height of fashion now with the young men so people might
just think that Ballylinan is full of very cool people. After the
first show, they shaved off their beards for charity. Some
£3,000 (more than €3,800) was raised and used to provide eye
tests and follow-up treatment in Africa. We've continued
with that tradition of beard-shaving and we've helped other
charities since.

When we were preparing for the last Passion Play, in 2015,
people said we'd never get the crowd, that things had changed.
But God, they were as anxious then to see it as they were
when it started nearly twenty-five years ago. We've had all
sorts of people playing Christ over the years – an electrician,
a farmer, a travel agent, a postman. We don't discriminate. A
local sergeant played King Herod on all occasions, and a very
good job he did too.

DJ has taken part every year and he usually plays the part
of one of the key soldiers.

Anna Marie has acted as Mary Magdalene a couple of times,
and she was lucky in 2010 that the garments worn were quite
loose as she was six months pregnant during the show. Her
son, Saran, and DJ's daughter, Dearbhla, were part of the cast
in 2015.

We've had bishops in the audience and visitors from the
US and the UK. I remember meeting a Swiss journalist who
was in Dublin and heard about the play and came down to
see it. She gave it a great write-up when she returned to Swit-
zerland. She said the characters in the story were no longer
figures from the Bible but were as real as if she was in their

company, journeying to Jerusalem with them. We've never had anyone from Oberammergau over to see our version, at least not that we know about, although they might have sent someone incognito to check up on the competition.

I couldn't say enough about the Passion Play. It is truly one of the most wonderful things in my life, after meeting John. People who have taken part would tell you that it has made the Easter ceremonies much more meaningful for them. And the best thing of all is that people who have seen our play, and seen the original one in Oberammergau, said our one was better. The German one can seat an audience of more than 4,700, so you don't get as close to the stage as you do in Ballylinan. Also, the Oberammergau one is in German, of course, which makes it harder to follow if you don't speak the language.

The Passion Play is quite a big logistical feat to carry off and given that myself and Mary Brennan were so involved, it made sense to use the NPA as the ticket office as well. That was fun, I can tell you. For every one call the girls would get for a stand at the Ploughing, they would get two for seats at the Passion Play – try figure that one out. But, fair play to the NPA, the directors have never questioned me when I do something like that. I suppose they see it as a good deed from the NPA, and maybe it puts us a little bit higher in the pecking order when the man above is doling out September weather. We can only hope.

All going well, we'll be putting on the Passion Play again in 2020, and it can't come quick enough. So if you see some very long beards going around Laois in a few years, you'll know the reason why.

15. We all have to do our bit

Canon John Hayes was ahead of himself when he founded the community organization Muintir na Tíre in 1937. The objectives he drew up eighty years ago are still as relevant today, or maybe even more relevant. Community spirit, neighbourliness, self-help – these are the things that will keep rural Ireland alive, and it's important that we don't forget them. You used to see it in action all the time, years ago. If a woman was widowed on a farm and she had a young family to look after, all the local farmers would help out, sowing the crop and helping with the harvest. Similarly, if a man was widowed, the local women would do what they could to help out, in terms of cooking and cleaning where necessary. That was part of life then and it's one of the things that I treasure about living in rural Ireland.

Nothing compares with the quality of life in the countryside – I couldn't ever imagine living in a city or big town. It's such a good way of life in a rural community and I think we have to preserve it for our children and their children's children. I always loved home and no matter where I've travelled in the world, the nicest part of going away is undoubtedly the coming home. It's not that I didn't enjoy my travels. I really do enjoy the experiences and the comradeship with the ploughing folk, but there is nothing sweeter than turning into your own driveway after being away for a while.

I've counted myself lucky to have been raised in a place like County Laois and I'll do all I can to make sure that we give the generations coming behind us the same opportunity to live in the countryside. There's nothing like growing your own lettuce, collecting your own eggs from the hen house and having room for a dog, if you want one. It's a wonderful environment for raising children.

Some of the things Canon Hayes said about invigorating rural Ireland should be repeated today. I have a press clipping from the *Irish Independent*, dated 30 June 1956, when he said it was 'high time for our government to realize that when a country had solved her rural problems, she had solved all her problems'. He talked about the tendency to put agriculture in a backward position while promoting industrialism, yet the real wealth in a country lay in its land. This was exactly the mantra of people like President Michael D. Higgins when the recession hit not that long ago. I remember him talking about the need to challenge the idea that progress had to involve urbanization, and how sustainable rural development would help the entire country.

I suppose you could say that supporting rural Ireland is one of my missions. We've had several invitations to move our office to Dublin, but there's no way I would consider it. At one time the NPA council was looking seriously at moving the office, but I wasn't for budging. I said: 'You can move the office, but I'm going nowhere.' I think we must keep the headquarters, and our identity, in the heart of the country. And we truly are in the heart of the country. Unless it was signposted, you wouldn't find us. And that's no harm. I think we wouldn't get as much work done if we were in the city, when you think of the commuting, and the distractions.

We came to another crossroads when Esso asked to be title sponsor to the Ploughing back in the mid-1990s. It would have been called the Esso National Ploughing Championships.

Now, Esso was a great sponsor and we could have done with the money, but I wouldn't even consider it because we would have lost some control over the event and I knew our Association wouldn't entertain that prospect. I remember telling one of our great supporters afterwards and he said: 'If you had accepted that, I would never have put a spade in the ground again for the Ploughing.' He was adamant that we keep our own identity and our own control over the event. It would have been an easy way out at the time, but what would have become of the Ploughing?

It's our duty to safeguard the future of the NPA. We have had to make some tough decisions at times, but one thing always guides us – will this be good for the Ploughing?

About ten years ago, a promoter came to us with a plan to take over the running of the Championships. They would look after everything and pay us a percentage of the earnings. We didn't contemplate that for a second because we would have lost everything. We would have had no control over the Ploughing if we had agreed. Our founder, JJ Bergin, would have been spinning in his grave. And we would have had to let our staff go because the promoter would have had their own people. I actually don't think it would have worked for the promoter either because all that goodwill and voluntary work would have been lost.

We had another approach, this time to run music events in the evenings on the Ploughing site, but we felt this would take us into a whole other sphere and it wasn't a direction we wished to go in. We just weren't comfortable with it. Our theory was that we should stick to what we were doing and work towards making it better every year.

There was another occasion when we had a really good guy who was released by his employer to do a specific job for the Ploughing. He had been doing the work for a few years and he was excellent, but shortly after one of the Championships, his boss came to me and said that unless we took on his company to do all the work in a certain area, he wouldn't allow us to use that employee again. He said he'd give me time to think about it, but I told him there and then: 'No.' We were very sorry to lose that guy because he was very good at his job, but we couldn't be held to ransom like that.

It's my firm belief that you cannot have the Ploughing in a locality without putting something back into the community. Take Ratheniska, for example, where we had the Ploughing between 2013 and 2015. That village, near Stradbally, County Laois, only has a church and school, but we were able to galvanize seventeen local sports clubs to volunteer at the event. We don't pay them, but we give contributions to their clubs, which gives them a great boost. They would have to sell a lot of club lotto tickets to get the same money. It also brings the clubs together and makes people feel involved. It's not only the GAA clubs. We asked all sports clubs in the vicinity of Ratheniska to apply to volunteer at the event. We used local contractors, where possible, for the preparation and restoration of the site.

It was the same with Screggan, which is hosting the Ploughing for the second year in a row in 2017. We contributed to their new community centre in Mucklagh by renting it for all our organizing meetings. We also had their grounds surfaced and we allowed them to sell our catalogues, which raised almost €9,000 in commission. That was a good injection of cash for the community.

But the Ploughing is only once a year and it's not enough to keep people in business. When people say, 'How do we save rural Ireland?', I honestly don't know what to say because

I don't feel there is a single solution that would solve all our problems. But I do think we must do more than complain, when it comes to protecting our services in the countryside. We have to use those services. There's no point complaining when the shop is shut if you never set foot in it when it was open. But having said that, I fully understand why people opt for a supermarket where the produce is fresh because of the turnover and there's a much greater variety of products. How can a corner shop compete with that? They can't, and I suppose that's why most of them have not survived. But you'd hope that a number of corner shops will always survive because of the convenience they offer.

We must stand up and be counted if we want rural Ireland to survive and thrive. I strongly believe you should be involved in your local community if you want it to thrive. If you want something done, then get up and do it. Don't complain that no one else is doing it. I know it can be inconvenient at times when you're working and you have children to ferry here and there, but you get a great amount of pleasure out of being involved.

I'm reminded of the Audrey Hepburn quote when she said, if you ever need a helping hand, it's at the end of your arm, and as you get older, remember you have another hand. The first is to help yourself, the second is to help others.

I know from experience that the same people are called on again and again to help out. And they are already so busy because they are involved in other things. But it has always been in my nature to get stuck in to whatever is going on. I don't know where I got this from because most of my family wouldn't really be like that, apart from a few of us. I've mentioned how my sister Eileen is very involved with the Irish Draught Horse Breeders' Association and she's also involved in the Irish Farmers' Association in Carlow. My brother Oliver is another good organizer. He played for Ballylinan GAA for years and has been deeply involved in the club's activities. He's often found taking money on a gate or helping out however he can when there's an event on.

I just love organizing things. The idea of sitting back and letting someone else do it is not in me. And if we all took that attitude, where would we be? It drives me mad when people complain that the same people are running everything, yet they refuse to get involved themselves.

I was one of the people who started the Irish Countrywomen's Association (ICA) guild in Ballylinan, back in 1964. And I'm sorry to say that there are only two surviving founder members – myself and Bridie Kaye. Bridie was a stalwart – she was treasurer, secretary, president, everything, and then you'd see her out picking up litter on the side of the road for the Tidy Towns Association.

We decided to set up our own ICA guild in Ballylinan because we were cycling in and out to the guild in Athy and we saw the demand for it in our own village. I took the job as president – I didn't want to, but pressure was brought to bear and I said yes.

Back then there were not many outlets for women in the countryside. They were basically tied to the kitchen sink and the farmyard. But this gave us a chance to get out and meet people. We had monthly meetings, including a social half-hour. We'd have talks and demonstrations and you'd come home inspired by what someone had said. You'd also come home full of local gossip and with the feeling that you were truly getting to know the women in the area. The meetings would put you in good form. I'm still a member and I wish I could attend more events, but time is always a problem. I regularly miss ICA meetings because I have to attend to Ploughing matters. But in fairness to our Guild president, Breda Hovenden, she will always leave a message with the girls to remind me of a meeting, cake sale or demonstration. When I give myself the time to go, I always love it.

I could be out every night of the week if I went to all the things I am invited to. I have eaten more than my fair share of cold dinners, abandoned hundreds of cups of tea and missed many news broadcasts because the phone rings just as I am sitting down. Whenever someone in the family is planning an event now, they would often check if I'm around before they settle on a date. Isn't it great that at my age I have a busier social life than the young ones?

Even though I miss so many of their get-togethers, I still admire the ICA, and I'm sorry to see that it is struggling now. I was at an event lately and I noticed all the grey heads there. It's very hard to get young women involved in it. They lead such different lives from us. They are dropping children to crèches, working long hours and rushing home to get ready for the next day. They have no time or interest in going to an ICA meeting. That's the reality for the ICA today and it's a pity that their beautiful headquarters at An Grianán, in County Louth, are not better supported. Women can go there for very good courses, whether it's cooking or mechanics you're interested in. I suppose, like rural Ireland, the ICA needs to look ahead and see what it can do to safeguard its future. And I would say to countrywomen, use your local ICA group before you lose it. There's no point in crying after it when it's gone.

The ICA has been very good to the Ploughing and I have even cajoled some of its members to help out at the Ploughing. Breda Hovenden, Carmel Brennan and Anne Lacey are all great for looking after our VIPs. Another staunch ICA member, my sister-in-law Anna Brennan, prepares the bandstand and never forgets to bring the scissors for the cutting of the ribbon which marks the beginning of the Championships.

I'm sad to see that we have lost many of the things that were good for community spirit. The Corpus Christi procession through the village was such a big thing when I was a youngster. The Blessed Sacrament was carried from the church and up the street, under a canopy, and the children who made their First Communion dressed up in their outfits. The teachers would take the petals from the peony roses and the children would spread the petals at their feet as they walked up the street. All the shopkeepers made a great effort with statues and altars and flowers in their windows. It was absolutely lovely, but then it was done away with, along with so many other things. I suppose we can blame ourselves for letting those things go. Some parishes didn't let the Corpus Christi procession go and I admire them for that. I'd love to see it coming back in our parish and I'd say there would be plenty of support for it. You don't have to be a holy Mary to enjoy the spectacle. The younger generation have totally missed out on that.

I still think you cannot beat the chit-chat outside a country chapel on a Sunday. It's absolutely wonderful. In some places, I've seen the men go to one corner to talk and the women go to another corner. You'll always come home with a bit of news. With the fall-off in Mass attendance, I often think that young people are missing out on this form of community bonding. But I suppose other gatherings will continue to bring people together, things like GAA and soccer matches, drama groups, whist and bingo nights.

The Annual Goose Club was a big community event for

us, too. This was organized to raise funds for the parish. Every parishioner would give a prize and they were raffled during the function. You could win a goose or a turkey or a hamper, and you'd be absolutely thrilled if you did. You'd always get the men coming in from the pub and buying tickets so that they might win a bottle of whiskey. We kept geese at home and we'd always give a goose to the raffle. That reminds me of the time John sold two geese to the doctor. He asked Anna Marie to deliver them. A few months later he remarked at the dinner table that the doctor had never paid for the geese. 'Oh he did,' said the bould Anna Marie. 'But sure I went shopping with the money.'

I'm sorry that we have lost so many of those things that made us unique. They brought people of all ages together and that's always important for community spirit.

I also feel great regret when I think of all the services we've lost in rural Ireland. The creameries were such a hub. Farmers didn't just deliver their milk and go. They met neighbours. They had a chat. Maybe there was a shop attached to the creamery and they'd go in and get a few groceries. It was very sociable for them and without that, perhaps some people could go for a week without speaking to another soul. The post offices fulfil that role now and we need to protect them. Everyone is entitled to basic services, such as a post office and a transport service, whether they live on a hill in Kerry or in the middle of Dublin. In fact, the person living alone on a hill in Kerry might rely on that post office far more than the person in the centre of Dublin who does everything by e-mail and pays all the bills online. Some-times decisions have to be based on more than the financial cost of something. The good of society must come into it.

Now, I have to acknowledge that you can't stand in the

way of progress. I'm not asking that the creameries be re-instated. I know that having the milk collected in the farm-yards is labour-saving instead of people bringing a milk can on a horse and cart. But still, there comes a point when you have to shout 'Stop' and not let any more services go. If we don't raise our voices, what will be left? And if we lose the people in the countryside, we lose the heart of it. Tourists love to see animals grazing in our well-kept countryside, but if our policies force everyone into towns and cities, tourists won't be flocking here to look at empty fields and abandoned homes.

There is an awful lot of talk now about environmental issues and climate change, but I think if we were eating more non-processed foods, like we did in my day, then we would be cutting out a major cause of environmental pollution. It's great to see children being educated on separating plastics, caring for the environment and thinking about their carbon footprint. There has to be a balance too, however. When the suggestion is made to downscale the number of cattle we produce because of the methane gas issue, I always pose the question: Will that reduce the world's carbon footprint by even a decimal place? I don't think so, but I tell you what it would do: it would reduce the number of Irish farmers even more. And that will have a knock-on effect right throughout rural Ireland.

The provision of broadband is still a huge issue for people in rural Ireland. There have been promises to resolve the problem for a long time, but we've yet to see any progress. This is a major factor for people trying to run small businesses in the countryside. You're trying to download large files or Skype someone abroad and the connection just goes. It's not good enough in this day and age. I know people who would work from home a lot more if they had the facilities, but they just don't have reliable enough broadband to allow them to do so.

Anna Marie divides her time between Ballylinan and Wexford, where her husband Declan is from, but she can't do any serious work involving broadband when she's in Wexford because the Wi-Fi is so bad. And she really needs the Wi-Fi connection, as secretary to the World Ploughing Organisation. I see her here, Skyping people around the world and having three-hour meetings with the WPO executive while sitting at her desk. Our broadband is okay, until it gets very busy and then everything slows down considerably.

If we want to keep rural Ireland thriving, we must ensure that fast, reliable broadband is brought to every corner of the country, right down to that farmer on the side of a hill in Kerry. We have so many great little businesses trying to get off the ground in rural Ireland and they just need that helping hand to really take off. The growing popularity of artisan foods is fantastic and it is definitely a story that we need to be shouting from the rooftops. We have such unique produce here that should be at the front of the shelf in international markets. Supporting those food producers is certainly one way that we can keep rural Ireland thriving because they also have a knock-on effect then for local employment and service providers.

I was happy to see the Government coming up with its National Framework Plan 2040, which is trying to prepare for the type of country Ireland will be in twenty years' time. With as many as one million extra people expected to be living in the Republic by then, now is the time to be planning ahead for housing, schools, transport and jobs. It was the lack of planning ahead that created our big broadband

problem, so it's encouraging to see the Government thinking ahead now. I was delighted to see our former Minister
for Agriculture Simon Coveney spearheading the plan, with
his vast experience of the countryside. It has now been taken
over by Eoghan Murphy, Minister for Housing, Planning
and Local Government, and I hope he won't forget about
rural Ireland in his work.

Brexit is another big challenge facing rural Ireland because
we rely so much on a thriving farming industry and the
United Kingdom is such an important customer for our beef
and other foodstuffs. After all, farming, along with tourism,
were the two bright lights during the recession when the
construction sector collapsed and shops shut down in every
town. Just under 10 per cent of our exhibitors at the Ploughing come from the UK and while we have no figures, I do
know that a substantial number of visitors to the Ploughing
come from the UK every year. We would hate to see that
changing.

We can't deny that things have changed in rural Ireland.
Three-quarters of the population of Ireland now live in
urban areas, which is a complete turnaround from the time
when I was growing up. I think we need to do a lot of work
to educate urban people on the importance of the farming
industry. People often think that because you have land, you
are rich. Well, I can assure you there are many farming families that may be asset rich but cash poor, and that is a terrible
place to be. At least there is now social welfare support for
farm families with very low incomes, but I don't think it's
the full answer. All a farmer wants is a fair price for his or
her product. How can it be right that a dairy farmer today is
paid an average of 32 cents per litre of milk produced when I
buy that same litre of milk in my local shop for €1.20? I can

appreciate that there needs to be a mark-up along the way, but it has to be within reason.

Then take the grain farmer: he delivers his load of grain to the local merchant but doesn't know what price he is going to get until the merchant decides it, based on market prices. Would you sell a product to someone and let them decide what they were going to give you once they had the goods in their hand? The farmer has no choice. I often remember John saying: 'Sure, how do they expect us to survive at all?' So the next time you hear a farmer complaining, just remember he usually has good reason – he only wants fairness in the marketplace.

The rural organization Agri Aware is great for highlighting the role of farming and the food industry and the NPA is glad to support the organization. It was founded in 1996 with a mission to improve the image and understanding of farming and the food industry among the general public. I think that's a very necessary thing as people have lost the connection between the food they eat and where it comes from. Along with nearly eighty other agri-related bodies and companies, the National Ploughing Association is a patron of the charity and we all fund its activities. One of its biggest successes is the family farm at Dublin Zoo. Children love to get a closer look at the animals and see how a cow is milked.

I think the lack of support for people working in rural areas will destroy our rural communities if it is allowed to continue. We need policies to encourage people to live and work in the countryside. Of course, people need more than farming to stay in the countryside, but I thought Michael D. Higgins made an excellent point in his wonderful speech at last year's Ploughing when he said there could be no vibrant rural communities without thriving family farms. It's so true. He said society must ensure that farming men and women are enabled to carry out one of the most important and beautiful human activities on Earth – the tending of the land and the cultivation of its fruits. Now isn't that really what it's all about?

16. A few life lessons

As I've been in this business for so long, people often ask me for advice on matters. The National Ploughing Association is seen as an independent, non-political organization and we have good relationships with the other rural organizations, so I suppose people feel that we have no agenda or bias. By no means do I think that I know it all – I'm still learning every day – but here are a few things I've learned that have helped to guide me along the way.

Never forget where you came from

You have to keep your feet well on the ground and never forget where you came from and the people who helped you to get where you are. And that goes for everyone in life, no matter what they do. I continually tell myself that I must not forget the grassroots of the NPA – the volunteers in our county ploughing associations. They get no recognition for the work they do all year round, and then when they come to the National Ploughing Championships we have hardly time to talk to them because it's so busy. That's why I always make a point of going to as many county ploughing matches as possible. I want people to know that we truly appreciate the work they do. I'm blue in the face saying it, but without the grassroots, we'd be nothing. If we forget that, it's the rock we will perish on.

I have surrounded myself with good people and that's why the Ploughing Championships work so well. From the man sweeping the road to the fellow who gets up on the bandstand, dressed to kill, every one of them is important and deserves recognition for what they do. Even though we have a large number of contractors at the Ploughing, it's the old faithfuls like Maureen Healion and Colette Gorry who we still turn to when we have to coordinate feeding the masses of volunteers. They've been doing this since they first got involved, when the Championships were down the road from them in Tullamore in the 1980s. And as the Ploughing expanded, so did the team, with Teresa Mahon, Lil Tracey and Helen Hally all rolling up their sleeves and getting stuck in. They have become part of the Ploughing family and go on tour with us every September for a week of hard work, a dollop of hardship and a good few laughs along the way. At the end of the day, it's all about the great sense of comradeship and being part of something special. If Anna May McHugh was the only one running the show, we'd have no Ploughing.

Nor would we have the Ploughing without our wonderful competitors. It all started with horse ploughing, and there is still such a strong tradition of horse ploughing in west Cork, Kerry, Louth and Galway, in particular. It's only when you go to their local matches that you see the dedication of the competitors and the journeys that they travel to come to the National Ploughing Championships. It's a similar case with Donegal competitors. My God, some of them are so far west in Donegal that the next parish is America. But wherever they may live, our people are never so far away that we'd forget about them. We would do that at our peril.

Never sign anything without reading it first

This was always the mantra of the co-founder of the NPA, JJ Bergin, and it's something I've never forgotten. I deal with all the contracts in the Ploughing, so it's something I am very conscious of. You have to read every word before you put your name to anything, but I know many people don't do this. I'm always particularly cautious when it comes to insurance documents. It's only when you have a claim that you discover you weren't covered because of a clause in some section that you didn't notice when you signed the policy. People can be very blasé when they sign these documents, but they are taking a big risk.

I am also extremely careful about what I put in writing. Once it's written down, it can be waved in your face at a later date if things don't go as expected. Ask yourself: could I stand over what I'm about to write in six months' or a year's time?

Keep your head when everyone else is losing theirs

This is something I repeatedly tell myself because we face many little crises during the Ploughing. Happily, the public doesn't know about most of them because they happen behind the scenes, and we always try to deal with the problems swiftly and decisively.

If a crisis situation arises, I make a point of keeping calm. When you are running an event the size of the Ploughing, and endeavouring to please the patrons, the exhibitors, the competitors and the landowners, it's essential that you remain calm and keep total control, with the help of your directors. Now inwardly my heart might be turning somersaults, but I cannot let that show. If people saw me panicking, how would that help the situation? They would probably say: 'Well if Anna May is losing the head, then we are all finished.' That's the time to show strength and calmness.

I feel it's very important not to put stress on other people around me. If, say, a car park is in deep trouble because of the weather, or there has been an accident, I just take a deep breath and ask myself: 'How can we deal with this and sort it as quickly and as effectively as possible?' Of course, preparation is the key to that. That is very true. If we have all the correct procedures in place, an accident will be swiftly dealt with by our medical centre and the log-jammed car park will be alleviated because we will have the vehicles ready to tow the cars out, or the panels to lay down. Then our stand-by car parks will be called into use. You have to be prepared for all eventualities. I'm a great believer in the expression: 'If you fail to prepare, you must prepare to fail.' Anything could happen when you bring tens of thousands of people together in a rural setting. And that's before you consider the Irish weather. I definitely do not believe in winging it. It's my duty to do everything I physically can to make sure that the Ploughing works well for everyone. And no one takes it to heart more than me when something goes wrong. I just want to send everyone away with a good flavour in their mouths, so I take it very personally if someone hasn't had a good experience. There are a few things you always must be prepared for in this business: be able to make quick decisions when necessary; be observant at all times; understand that you are not in a nine-to-five business; and always be ready to improvise.

Never ignore a problem

Sometimes it's tempting to do nothing, when you don't know what to do. Or to try to cover up a problem and hope no one will notice. I would always caution against that because these things come back to haunt you. I'm thinking of the time we announced the wrong result on the last day of the Ploughing in Ardfert, in 1984. There was a lot riding on it because the winner would go to the World Ploughing Contest. Anyway, somewhere along the line, the order of the competitors and the plots got mixed up and the wrong man was called out as the winner. He was a lovely man from Kilkenny and there was no one more surprised than him when his name was called. I was thinking of that again when I heard about the 2017 Oscars controversy when *La La Land* was mistakenly called out as the Best Picture winner, instead of *Moonlight*.

Unlike the Oscars controversy, we didn't realize the error at the time and I went back to the hotel, glad that another Ploughing had passed off without incident. It was about midnight when I got a knock on the door. Questions had been asked about the result and colleagues had been going through the sheets. They realized that the order had been mixed up and the wrong winner had been announced. My heart sank, but I knew we had to deal with the situation. And we did that, first thing in the morning.

The two competitors in question were tracked down as quickly as possible and, in fairness to them, they were very understanding. That made the whole incident more manageable. But some other people were very annoyed, and some of them were trying to blow it up into something bigger. I even remember one lad calling for my resignation at the time. But what more could we do? We had corrected the mistake, apologized and written to the county ploughing associations. Everything was smoothed out in the end. But it shows that you cannot be ready for every eventuality. Something will always come along that you haven't dealt with before and you have to make the best decision at the time. It's easy to be a wise owl afterwards, when you were not the person faced with the decision at the time.

I remember someone saying to me that we could have swept it under the carpet and no one would be any the wiser. I suppose we could have done that, but it wouldn't be fair on the competitors. And we would have to live with the knowledge that we had covered up a mistake and caused someone to miss the opportunity of representing this country abroad. We certainly learned our lesson from that occasion and nothing of the sort ever happened again. I do not believe in smoothing over the cracks. I want to find out where the cracks are and how they can be mended for good.

Sleep on things

As I mentioned earlier, this was always great advice from my late husband, John, and it's something I would always emphasize. It's very tempting to charge in, all guns blazing, if you are annoyed about something. And as a red-head, my first instinct is to fire on all cylinders, shoot first and ask questions later. But of course, that can often be the worst thing to do. Take it home, mull it over and sleep on it. You'd be amazed how different things look in the morning. The major offence you thought someone might have committed against you suddenly seems insignificant. But if you'd acted when it happened, the matter might have escalated into something far worse.

Trust people unless they give you a reason not to

I would be inclined to trust people and take their word that they would do something, but I see a big difference between me and Anna Marie. She's from a different generation and she would want everything in writing, signed and sealed. A handshake wouldn't do. I suppose that's the way the world has moved.

Now, if someone lets you down, that's a different story. I would be very hurt indeed if people I trusted let me down. As I said earlier, I am very loyal to people and I expect that loyalty to be reciprocated. Based on my experience, I think I have fairly good instincts when it comes to reading some-one's character. You can feel it when you meet someone if you are going to work well together. I always strive for good working relationships and if I sense that people are not work-ing well together, I would change things around, maybe move one of them out of that situation. If there's friction, the attention is not on the work that needs to be done, so har-mony is essential.

Everyone is good for something and I like to think that one of my talents is detecting that ability in someone and then put-ting it to work. When that person gets the opportunity, they would surprise you beyond belief. Some people would never get the chance to even speak at a meeting, and I would try to bring those people into the discussion. I know myself how much I hate going out of a meeting if I haven't made some kind of contribution. You feel a bit useless otherwise.

Think twice about giving a second chance

We have only one event a year and we cannot afford to have weak links in the chain. So, if someone lets us down, it would be very unusual that we would give them a chance to let us down again. There would want to be a very good reason for the mistake they made. We do expect a lot from people, and I think we expect more with every year that passes. I expect as much from our unpaid volunteers as I do from our paid employees. And why not? If a volunteer manning a car park is rude to a family, it might turn that family off ever returning to the Ploughing. They see that volunteer as the face of the National Ploughing Association and the rude attendant has succeeded in planting the idea in the family's head that the NPA is not appreciative of their custom. They don't care if the attendant is a volunteer or a paid employee.

And believe me, we always hear about it if someone isn't treated well. We have representatives from every county on the NPA council and they let us know very quickly if their neighbours or relations have had a bad experience. Even if we've just held the most successful Ploughing ever, it will be the negatives we'll hear first at the next meeting. It won't be the congratulations. It could be as small as someone thinking the bread was a bit hard in their ham roll. There's no danger of getting too big for your boots in this job. But that's the way I like it because it keeps me keen and alert. It's never good to rest on your laurels.

Meet people face-to-face

I'm probably old-fashioned about this, but I'm a great believer in meeting people face-to-face, rather than communicating by e-mail or text. I'm not an e-mail person at all. I know it's a brilliant method of communication and a very quick way of getting answers to questions, but for the real hard business side of it, there's nothing better than meeting face-to-face. When I'm doing contracts, I love to sit down with people and talk them through it and see if there's anything they're unhappy with. You know exactly where you stand when you meet someone because you can sense what they are really like, and how genuine they are. You'd seldom get that from an e-mail.

I think you make far more progress through face-to-face meetings than pinging e-mails back and forth all day. There's another problem with e-mails. Because it's so instant, you might be tempted to fire off a cross e-mail without thinking. Once it's gone, you can't reach into the computer and take it back.

I'm not a lover of texting either. I'm old-fashioned that way, but I've managed okay so far and I see no reason to change. I use a mobile phone all right and find it great for security when I'm in the car on a long journey. You never feel alone when you have the phone with you. Mind you, the mobile can be a source of great annoyance if you give your number to too many people. My number is like the Third Secret of Fatima. Very, very few people have it and I'm determined to keep it that way. Otherwise, you'd never be off the phone.

Treat people well

I feel it's very important to treat people well in business. As well as it being the right thing to do, ethically, it also makes business sense. If you don't treat them right on your first encounter, you can't return to them because there's bad feeling there. So, you might do well financially out of that first encounter, but it could cost you more in the long run because you've lost a good business contact. I always pride myself on knowing that I could return to any one of the farmers who hosted our ploughing events and we could do business again.

Apart from all that, we live in a small country, and if we didn't treat people well, we'd have no Ploughing Championships because we'd eventually run out of people to do business with.

Think before you speak, especially if you're talking to a journalist

I'll have every journalist after me about this one! But they know exactly what I'm talking about. I learned this the hard way, early on in my work with the Ploughing. I was being interviewed by a journalist and something came up about the Northern Ireland Ploughing Association. Relations were not brilliant between our Association and the Northern Ireland Association in the early days, but we all get on very well now and we are a huge support to each other. I look forward to the Northern Ireland Championships every year, which gets a great attendance from the Republic of Ireland. Anyway, I cannot remember exactly what I said, but it was just one throwaway comment that I put no value on.

I didn't know then that journalists had the skills to build a big story out of very little. This big story appeared in one of the evening papers and it landed me in a lot of bother at the time, so I learned a valuable lesson. Consider every word you are going to say before you open your mouth. Journalists have a way of drawing you in and then they will elaborate very much on that one thoughtless comment. Suddenly you open the paper and see a big article with a controversial headline all built around a little afterthought you might have voiced when you were walking out the door.

Speaking of Northern Ireland, we always made a point of sending competitors to their ploughing competitions, even at the height of the Troubles. There was always a fear about our safety when travelling across the Border. The soldiers would stop you for questioning and maybe search the car and you'd be so nervous, even though you had nothing to hide. But we felt that we had a duty to support our colleagues in Northern Ireland when they were going through such hard times. It's all part of treating people well, and I'm glad that we took that stand. We have good friends in Northern Ireland and we have the height of respect for our Northern Ireland colleagues.

Share your worries

I could do with some advice on dealing with worrying myself because I am an awful worrier. I get very anxious about things and I just cannot switch off from work. I'd love to be a person who could close the office door and forget about work, but that's just not me. I live the Ploughing every minute of every day. There are a few things that help me to deal with worries, though. I make a conscious effort not to sit at home brooding about things. Get out of the house and meet people and that can take your mind off the worry. And share the problem. I always find that it eases the anxiety if you discuss the issue with someone. They mightn't have a solution, but it still takes a weight off your shoulders to share that worry.

Family is everything

The Brennan family was always very united, and that's something we got from my parents. Hand on my heart, I cannot ever recall a serious row or falling-out between the eight of us. My father put a huge emphasis on supporting each other, probably because he was an only child. He repeatedly said to us: 'Wherever you are, stay united and help each other.' That advice stuck and whenever anything happens to someone in the family, we all rally around. My sisters-in-law and brothers-in-law are the same. We would have always congregated for family occasions and if anyone was missing, my father would want to know why. I know it's hard work bringing all the family together sometimes, but it's worth it all to see young cousins playing with each other and building on the family memories. I would hope that DJ and Anna Marie and their children would continue that sense of family unity.

It's so sad to see family arguments splitting people up. I know many families have a difficult sibling or two, but I think you must try to work with them and include them. Family is family and sometimes that contrary individual might just need an extra bit of nurturing. Give it to them. What do you have to lose? You don't want to be sorry when they're gone and you find yourself wishing that you'd done more to keep them in the fold.

Keep the faith

As I've said before, I'm not excessively religious by any means, but I would love to think that the generations follow-ing me would keep the faith and hand it down the line. Prayer can be a great support and religion gives you a set of morals to live by. I'd never criticize anyone's beliefs, or the lack of them, but I do think it helps to have religion in your life. It has definitely helped me through good times and bad, and God has given me great guidance on so many occasions. I often wondered how I would manage if John died before me and I prayed that God would give me the strength to handle it. As you know, he did die before me and I believe God got me through that difficult time. I still feel very close to John – I really feel that he is near me – and it gives me great comfort to think that we will be reunited in heaven someday.

Be kind

I think kindness is one of the more underrated virtues. It costs nothing to be kind and considerate and if everyone was kind, think of the world we'd have. I suppose I would have been considered a jolly individual when I was growing up and I have a naturally happy disposition. I always try to smile at people. A smile costs nothing and can give someone else a boost. You can't go around with a sour face all the time. I always make a point of saying 'Good morning' to people on the street, and sometimes you surprise people into smiling and talking to you. If they don't have a smile for you, then give them your smile.

When I did the *Meaning of Life* programme, I had to think in advance about what I believed the meaning of life was. I came to the conclusion that it was about living a good, ful-filled life, doing good where you can and helping people as you pass through life. That way, your living will not be in vain, as that old gospel hymn 'If I Can Help Somebody', goes One of my favourite singers, Josef Locke, did a lovely ver-sion of that song, many moons ago.

I think it's lovely when men hold doors open for women, or young lads give older people their seats on buses or trains. And I think it would be wrong if you didn't accept a kind offer like that. There is still a lot of kindness in people. You see it when someone dies. My God, the heaps of sandwiches and cakes that arrive at a house for the wake. People can be very good.

On the other hand, people can be quite cruel, especially when they are online. They maintain that they are entitled to express their opinion without any consideration for other people's feelings. County councillors and politicians some-times get a very hard time. I suppose we need a figurehead to blame for all our problems, but politicians are often restricted in what they can do. Of course, they'll promise the sun, moon and stars when they are out electioneering, but when they get into office, reality kicks in when they try to do some-thing. Politicians have a tough life, with long working hours, lots of travel, high expectations and little thanks. It's a life I would never want to have.

We are living in a more demanding society now than years ago and people have very little patience if you do not live up to their expectations. Every complaint goes online in seconds and is magnified. But I would urge people who are about to tear strips off someone for a perceived failing to just remember that they too could be in that situation someday. Being cooperative is always the best way to a positive result.

Best piece of advice I have received

Someone asked me recently about the best piece of advice I ever got and I immediately thought of the line I read on a calendar once: What you think about yourself is much more important than what others think of you. I thought that was very apt, although I know it can be hard at times to ignore what other people think. But as you go through life, I think you do become better at doing that. Wisdom comes with age, and a wise person learns something every day. But of course there are people who will never learn because they are not open to different points of view. I try to be open to advice and take something from it that will be helpful.

I've definitely taken on board the adage that you should surround yourself with people who know more than you. I've always tried to do that because we all have different gifts and there's always someone who is a better expert than you at a particular thing. Having all these talented people around you also keeps you on your toes and improves your own performance. I certainly believe that the day you stop learning is the day you die.

17. Looking forward

Now, I have a confession: I actually hate talking about myself. You might find that hard to believe if you've ploughed through the past sixteen chapters, but it's true. I suppose it's an Irish thing. You don't want people to think that you have a swelled head. No one likes someone who's always blowing their own trumpet.

So why did I write this book? It's out of shame and embarrassment really. You've no idea the number of times I've been asked to write a book. Publishers, journalists and ploughing people have asked me over and over again. I think they're afraid I'll go to meet my maker and much of the history of the National Ploughing Association will disappear with me.

Our own members would say: 'Oh for God's sake, just write something and put it aside so that we have the record. It would be a pity to lose the memories of how it started and how it grew.' And people would say to me: 'Would you not leave something so that your grandchildren and their children could read about your life?' I can assure you that I'm hale and hearty and have no intention of dying any time soon, but then again, that's probably the best time to tackle a project like this. So, when Penguin Books came to me in early 2017, I felt the time was right and I couldn't put it off any longer.

Putting this book together has forced me to take stock and look back on my life, and I must say that it has been a very good one. It was fate that brought me to JJ Bergin's to do a bit of typing that day in 1951, and everything just seemed to fall into place after that. Now, every day wasn't a rosy day in the Ploughing. I had my problems too, and I tried to solve them in the best way possible.

Sometimes when I'm looking at the map of the latest Ploughing site, or taking on more staff, I marvel at how far we've come since the days when it was just JJ Bergin, me, and his little school case. I have been very fortunate with staff and in all my years as managing director, I've never had to let someone go. I would imagine I'm in a minority on that score. Maybe it's the nature of the work, or the fact that the NPA is a voluntary organization, but the team always puts its best foot forward and thinks nothing of working well into the night, seven days a week, in the weeks leading up to the Ploughing. As I've said so often, I'm only part of a team, and what a wonderful team we have. I'm very proud to have my brother JJ's daughter, Louise, working alongside me now. It's no surprise that she inherited her father's great instinct for the business of the Ploughing.

When the Ploughing comes, we have about 600 people working on it, but many of them are volunteers. Judges, stewards and supervisors all give their time for free to the Ploughing every year. That tradition has never waned. Volunteers love to offer their help and in fact they would be disappointed if they were not asked to come along. The ploughing associations in every county see this as their event, and indeed it is their event, and they would be insulted if they were not asked to pitch in. I have no doubt that this voluntary army is one of the main reasons for the success of the event.

As you'll know by now, there was never a masterplan to grow the Ploughing into what it is today. It was like climbing the steps of a ladder, taking each rung at a time. The important job now is to control the growth and maintain what we have. There's always the danger of a disastrous year or two, but we are prepared for that. All I can hope is that the Ploughing will continue to prosper long after I'm gone. I've slowly come to realize the truth in JJ Bergin's desire to leave a well-oiled machine behind. I must have been the machine he was leaving behind, but the oil is running very scarce at this stage in my life!

I feel I am well-respected for my experience in the NPA, and I hope that continues. I'm not foolish enough to think that everyone is singing my praises, and I know there are begrudgers out there, but sure you can't help that so there's no point worrying about them. I don't listen to that kind of stuff and if there's contrariness or backbiting, I pretend I don't see it.

I would love to think that I would have no enemies when I leave the organization, whenever that might be. And if I have, that the enmity has been created in someone else's mind. I've tried to be as good as I can to everyone. Now a few people have left the ploughing organization over the years, following differences of opinion, and I was sorry to see them going, but I hold no ill-feeling for them. I hope they feel the same about me now. If they don't, I'm sorry they feel that way. I know that I'm often depicted as the villain of the piece if someone has a falling-out with the NPA because I'm the face of the Association. But in truth I always try to be the peace-maker if there's a conflict because I hate to see people falling out. I suppose I must live with being depicted as the villain because that's one of the joys of being at the helm.

The Ploughing has been very good to me. I worked hard, to be sure, but I love working. I love the feeling of waking up knowing that I have a busy day ahead. I'd say I'm working as hard now as I was thirty years ago, and I can't see myself working any less. I always feel the day isn't long enough for all I have to do. I get up around 7.30 am and never go to bed before midnight. I remember Gay Byrne asking me on the *Meaning of Life* programme if I took a nap during the day. A nap! I was looking at him like he had two heads. He said lots of people my age take a nap, but it honestly wouldn't occur to me to go for a lie-down in the middle of the day. The girls in the office would think there was something wrong with me if I took to the bed in the middle of work.

Sure, if I go to a wedding, I'm definitely going to be one of the last to leave the dance floor. I have often been spotted in the residents' bar afterwards, sipping a mug of tea at 4.00 am. But even then, I will be one of the first down to breakfast. I suppose I'm just blessed with good health and great energy. Thanks be to God, I've never had a serious illness and I've only stayed in hospital three times – for the birth of my two children, and when I had my hip done. I wasn't kept in hospital overnight when I had that car crash in 2016. I attribute my good health to hard work and having a purpose in life. When you have a job to do, it keeps you alert, and working with young women in the office also keeps me focused.

One of the first things I wrote down in this book was my date of birth. I know my age has been a source of great speculation, particularly when anyone is writing an article about me. You can't tell your life story without giving your date of birth, so I had to bite the bullet. I haven't been forthcoming about my age in the past because, rightly or wrongly, I feel people judge you differently because of it. I do it myself all the time. If I hear of an accident involving an eighty-year-old

I might say, 'Sure, he shouldn't be driving at that age,' and then I catch myself and remember that I'm eighty-three years old and still driving hither and yon.

I would be afraid that when my age is revealed, people might say: 'What can she be doing at that age? How active is she really, or is she just a figurehead?' It's totally wrong and ageist, I know, and I shouldn't think like that, but I do. And many other people do too.

I have the height of respect for people working in our healthcare system, but I do feel that sometimes we tend to write off older people when it comes to their health. There might not be as much effort put into investigating the cause of the illness, or the best treatments, purely because of their age. I don't think young people would be treated in the same way.

I think we can do better for older people in hospital.

We all deserve respect, whatever our age. I love to see younger people helping older people across the street. I would always offer a hand to an older person if the occasion arose and I get great satisfaction out of helping older people. But here I am again, forgetting that I'm old. My sister Betty would often say to me, 'What will we do when we get old, Anna May?' And then she'd laugh and say, 'But sure we are old'. We still feel young. I always thought you'd feel old after a certain stage, but that feeling hasn't hit me yet. I hope it won't for a good while.

I think attitudes have to change with regard to the employ-ment of older people. You see prison wardens and nurses aged sixty-five being forced to retire and they have so much more to give, but they don't have a choice. And then it's very hard to get other work at that age. We are wasting all that experience and talent by forcing public servants to retire for no reason other than their date of birth. I always say it's not the years in your life that matter, but the life in your years that counts. I was delighted when President Michael D. Higgins quoted me saying that, when he was campaigning to become president.

People often ask me about retiring. If I ever got the feeling that I was going to be pushed aside, I would get out before that would happen. I would hate to be in a position where someone felt obliged to say, 'Look it, Anna May, you're not able to do it any longer. It's time to go.'

I'd hope that I would realize it was time to go before anyone else did. I hope I'd recognize when I wouldn't be able to do the work to the best of my ability. You must have your wits about you in this job because you will meet many obstacles. There's no stipulation in our company documents about a mandatory retirement age, but I must go before our AGM every year for re-election as managing director. All our officers must do the same.

At the moment, I feel I am entirely capable of meeting all the demands of the job. You need very good physical health in this job because you have to go into fields to see how suitable land is for ploughing and how suitable it is for exhibitions and machinery. You're always on your feet, always on the go, particularly in the last few months before the Ploughing.

Anna Marie is one of three assistant managing directors and people often say she's the obvious successor as managing director. But as I regularly say, the Ploughing isn't owned by the McHughs. I would never pressurize her to go down that road and it's not up to me anyway. That would be a matter for the NPA council.

People tell her it must be grand to be working with her mother, but I don't know if she'd agree. She often points out that when she goes on holidays and rings home, I never ask her if she's enjoying herself – I only talk about work. 'So and so rang, and what will we do about this issue, or did you do anything about that problem?' Of course, you wouldn't dream of doing that with a regular employee, but that's how it works with family members who work together.

Anyway, I hope my succession is not something that will arise anytime soon. To be truthful, I wouldn't like to retire just yet because I'm still enjoying it. I love meeting people, and you definitely meet all sorts in the Ploughing. If I had a cent for every person who asked me for a ticket to the Ploughing over the years, I would certainly be a millionaire by now. And it's not that it's expensive, or that tickets are scarce, like All-Ireland tickets. People just like the idea of getting something for nothing, I suppose. I make a point of paying my own way whenever I'm going to an event. It's not good to be under an obligation to anyone. Having said that, from time to time people present me with flowers or chocolates, or wine at Christmas. That's not much use to a Pioneer like me, but the bottles come in handy if someone is running a raffle or looking for spot prizes. And, to be honest, the minute I receive a beautiful bouquet, I can picture exactly where it would work well in the church. So, the person who presents it to me can rest assured that it is being enjoyed by very many people, and not just me.

If you asked me would I change anything about my life, I'd have to say no. In one sense the Ploughing ruled my life, but there wasn't one week when I said to myself, 'God, I'm sorry I took this job.' I could have had a far simpler life as a farmer's wife, but imagine all that I would have missed. I would have missed some rare experiences, that's for sure. How many people get an opportunity to travel the world as I have done, and am still doing? I was so lucky. We're going to the World Ploughing Contest in Kenya in December 2017, please God, and next year it will be Germany, followed by the USA in 2019 and Russia in 2020. I'm hoping to be there for every one of them, if God spares me.

Through the Ploughing, I have received so many invitations to things I would never have been to. I have received so much praise and recognition for the job I did, but I have to say that the Association made me, rather than me making the Association. My husband John used to get a great kick out of people recognizing me if we were travelling around the country on our holidays. 'She'll be happy now,' he'd say, if someone mentioned the Ploughing to me.

It landed all sorts of opportunities in my lap. I remember the then Minister for Agriculture Joe Walsh ringing me in 2002 and asking me to sit on the board of Teagasc, the agriculture and food development body. When he called, he just said, 'This is Joe Walsh', and of course I never thought it might be the minister. I just thought it was someone inquiring about the Ploughing. He said again, 'This is Joe Walsh', so I said a bit impatiently, 'Yes, what can I help you with, Joe?' And then he identified himself as the minister. I was mortified.

Again, thanks to the Ploughing my home is full of the numerous awards I've won. They are a lovely reminder of so many special occasions. I won't bore you by listing them all, but a very memorable one happened in 1998 when I got a Rehab Person of the Year award for outstanding contribution to society. It was such a great honour to be recognized at that. I suppose it's different for people who live in Dublin or a big city and are probably used to being recognized for their work. But for someone like me, working away in the heart of the country, it's amazing to think that a selection committee came together and noted the work we'd been doing.

Bertie Ahern was taoiseach at the time and so he presented the award in the Burlington Hotel, in Dublin. I remember going to McElhinney's in Athboy to get a blue dress for it and being very excited to meet the other recipients. I was often sad that my mother and father were not around to see any of the awards being presented. They would have got a great kick out of it.

They would have been particularly proud to see me getting the Lifetime Achievement award at the *Farmers' Journal* Women and Agriculture awards in 2013. Like most farming families, the *Farmers' Journal* was like the Bible in our house, so I know it would have meant a lot to Mam and Dad.

I never got to university, so I certainly never dreamed that I would be conferred with a degree one day. Again, thanks to the Ploughing, Dublin Institute of Technology presented me with an honorary doctorate of philosophy in 2006, while NUI gave me an honorary doctorate of law in 2014. Better again, I got to keep the gown and tasselled cap when NUI honoured me. It's something I never imagined hanging in my wardrobe.

Sometimes I worry that people will say, 'Who does she think she is, getting all these awards?', but I would always stress that these awards recognize everyone in the Ploughing, not just me, and I really hope they feel a sense of pride in their Association.

Is there anyone I would like to meet? Well, I would love to meet Queen Elizabeth. Mary McAleese had such nice things to say about her when she came to Ireland. It was a very emotional visit for everyone and I think it was such an inspired idea to invite the Queen to Ireland. It's something the country will never forget, especially the way she wore green and spoke the few words of Irish. It was very touching, and I think she went up in the estimation of a lot of Irish people after the visit.

I would also like to think that I might meet another Pope before I go to meet my eternal reward. Perhaps with Pope Francis coming to Ireland next year, who knows?

Like many people of my age, my family is shrinking. Betty is sadly in the later stages of dementia as I write this, but Stannie, Oliver and Eileen are still going strong, and I'm sure Paddy, JJ and Gerald are looking down on us from heaven. It's lovely to see the new generations getting interested in ploughing. DJ and Fiona's children, Seán Óg (11), Tadhg (9) and Dearbhla (7), all have their ploughing overalls and I'd say they'll have a go at ploughing when they get older. I remember them coming into the house one day looking for judges' sheets. They marked out plots and were judging each other as they ploughed with their toy tractors. Anna Marie and Declan's son, Saran (7), also loves going ploughing and I would be amazed if some of his parents' passion for it doesn't rub off on him.

I'll never forget the day Saran was born. We were on our way to a Ploughing meeting when we stopped off at the Coombe Hospital for a check-up, this was about a month before he was due. The Ploughing meeting never happened because Anna Marie was taken in and they delivered the baby there and then. It was a very emotional time because we didn't know if he would be okay or not. I remember I was in the theatre with Anna Marie for the birth because Declan couldn't get up in time. I was holding her hand and the doctor was holding my hand and I think I needed the hand-holding more than Anna Marie. It was a huge privilege to be present for the birth of a grandchild and it's a memory I'll always hold dear.

I take great pleasure in the fact that DJ loved farming from a young age, and he works the farm here in Fallaghmore. It means I can be involved in farming when I want to be. After all, I'm still useful for standing in a gap when cattle are being moved. I like to think I inherited my mother's good eye for a sick animal, so it's good to be able to contribute when needed. I still keep a few hens and there's nothing nicer than collecting the eggs in the morning, still warm from the nest. It's very different from the old days when John and myself were filling bags of corn, bucket-feeding calves and rearing turkeys.

I still love to potter around in the garden and I'm at my absolute happiest in Fallaghmore, in the home that myself and John created. I would honestly much prefer a cuppa at my own table than a dinner in the fanciest Dublin restaurant, and I enjoy chatting to neighbours outside Mass just as much as I enjoy meeting celebrities. I've never craved the high life in any shape or form, so it amuses me when I end up in these situations. People often say to me, 'Sure you are a real celebrity, on the TV and meeting such important people. How do you come down to ground level at all?' In truth, it means more to me to get a note from an exhibitor I gave a helping hand to, or a positive result from advice I may have given to a ploughing club in a crisis. A nice memory shared with a relative of someone who gave years of their life to the Ploughing is more valuable than any meeting with a celebrity.

I have always felt a great affinity with country life and farming and feel very fortunate to have been born into such a lifestyle. Sometimes a little thing will happen that just stops you in your tracks. Take the time a very young lamb arrived at the back door late on a Christmas Eve. You might think that was not an extraordinary event on a farm, but if you knew the trek the lamb had to make, you might think differently. She had to get out of one lambing shed, find her way through two gateways and down a steep haggard, before arriving at the back door. I was dumbstruck when I saw the little creature standing there. It's at times like that when I think there's definitely a greater power at work here.

So, what's next? Who knows what lies ahead? I don't worry about death or think too much about it. I believe I've lived a good life and I've served God well. I've done my best to be peaceful with people and if they needed help, I tried to be at their door to give it. From time to time I hope I have helped get someone through a tough time, or over a hurdle that may have seemed insurmountable if they were attempting it alone.

Whenever death comes, we have no say in the matter anyway. I'd never plan the details of my funeral either. One day I said jokingly that I'd like my coffin to be laid out on the office desk. I remember my sister-in-law Mary Brennan coming back to me and quietly asking if she'd have to fulfil that wish. I had forgotten to tell her I was joking.

Do I think about life after the Ploughing? Not really. My two predecessors passed away while they were at the helm, so I guess that may be my destiny too, and what would be wrong with that? All I want to do is to work for as long as possible. I still love getting up in the morning to face the next challenge that the Ploughing, or life, presents. If I could be granted one wish, it would be that I would be blessed with good health to watch my grandchildren growing up healthy and happy.

I have been given so much in life that there's nothing more I could wish for.

I have been privileged to have worked with some of the most outstanding people I could ever have hoped to walk my path with me. There have been many occasions when a pillar of the Ploughing has passed away and I give a bit of a shudder and ask myself: 'Where will I go from here? That person was a great mentor.' But you know, I keep going, and I soon find that someone else has been there on the sideline all along, just waiting for the chance to get on the pitch and play their part. The NPA has benefited more than I could ever describe from the dedication of these members, past and present, from all over the country. I haven't been able to mention many of those people here, but I hope they know who they are.

I'd like to think I've used my abilities well, and I hope I will be remembered more for what I have done for others than for myself. As Muhammad Ali once said: 'Service to others is the rent you pay for your room here on Earth.' We are only custodians passing through, after all.

It truly has been a fulfilling life, a life well lived. And, you know, I think there are a few chapters yet to come.

God speed the plough.

Appendix 1: National Ploughing Championships

Date	Venue
16 February 1931	Athy, Co. Kildare
19 February 1932	Gorey, Co. Wexford
15 February 1933	Clondalkin, Co. Dublin
13 February 1934	Athenry, Co. Galway
13 February 1935	Mallow, Co. Cork
12 February 1936	Tullamore, Co. Offaly
9 February 1937	Greystones, Co. Wicklow
10 February 1938	Oak Park, Co. Carlow
8 February 1939	Killarney, Co. Kerry
24 January 1940	Thurles, Co. Tipperary (Junior classes)
1 February 1940	Co. Kilkenny (Senior classes)
12 February 1941	Co. Cork (Junior classes)
18 February 1941	Navan, Co. Meath (Senior classes)
12 February 1942	Cloghran, Co. Dublin
12 February 1943	Portlaoise, Co. Laois
9 February 1944	Ballinasloe, Co. Galway
21 February 1945	Tipperary town, Co. Tipperary
7 February 1946	Balbriggan, Co. Dublin
11 February 1947	Maynooth, Co. Kildare
19 February 1948	Limerick Psychiatric Hospital Grounds
10 February 1949	Drogheda, Co. Louth
9 February 1950	Bandon, Co. Cork

1 February 1951	Killane, Co. Wexford
7 February 1952	Athenry, Co. Galway
11 February 1953	Mullingar, Co. Westmeath
11 February 1954	Cahir, Co. Tipperary
10–11 February 1955	Athy, Co. Kildare
1–2 February 1956	Nenagh, Co. Tipperary
7–8 February 1957	Boyle, Co. Roscommon
12–13 February 1958	Tramore, Co. Waterford
28–29 January 1959	Burnchurch, Co. Kilkenny
11–12 November 1959	Oak Park, Co. Carlow
9–10 November 1960	New Ross, Co. Wexford
8–9 November 1961	Killarney, Co. Kerry
7–8 November 1962	Thurles, Co. Tipperary
6–7 November 1963	Athenry, Co. Galway
18–19 November 1964	Danesfort, Co. Kilkenny
17–18 November 1965	Enniskerry, Co. Wicklow
2–3 November 1966	Wellington Bridge, Co. Wexford
25–26 October 1967	Tullow, Co. Carlow
23–24 October 1968	Banteer, Co. Cork
22–23 October 1969	Rockwell College, Co. Tipperary
28–29 October 1970	Danesfort, Co. Kilkenny
27–28 October 1971	Finglas, Co. Dublin
25–26 October 1972	Rockwell College, Co. Tipperary
3–6 October 1973	Wellington Bridge, Co. Wexford
16–17 October 1974	Watergrasshill, Co. Cork
15–16 October 1975	Bennetsbridge, Co. Kilkenny
13–14 October 1976	Ferns, Co. Wexford
19–20 October 1977	Cashel, Co. Tipperary
11–12 October 1978	Knocktopher, Co. Kilkenny
10–11 October 1979	Watergrasshill, Co. Cork

7–8 October 1980	Rockwell College, Co. Tipperary
7–10 October 1981	Wellington Bridge, Co. Wexford
12–13 October 1982	Edenderry, Co. Offaly
5–6 October 1983	IDA Grounds, Waterford City
3–4 October 1984	Ardfert, Co. Kerry
2–3 October 1985	Kilkea, Co. Kildare
8–9 October 1986	Urlingford, Co. Kilkenny
7–8 October 1987	Tullamore, Co. Offaly
4–6 October 1988	Oak Park, Co. Carlow
10–12 October 1989	Oak Park, Co. Carlow
2–4 October 1990	Oak Park, Co. Carlow
8–10 October 1991	Crecora, Co. Limerick
29 September–1 October 1992	Carrigtwohill, Co. Cork
28–30 September 1993	Clonmel, Co. Tipperary
27–29 September 1994	Enniscorthy, Co. Wexford
26–28 September 1995	Ballacolla, Co. Laois
2–5 October 1996	Oak Park, Co. Carlow
30 September–2 October 1997	Birr, Co. Offaly
29 September–1 October 1998	Ferns, Co. Wexford
28–30 September 1999	Castletownroche, Co. Cork
26–28 September 2000	Ballacolla, Co. Laois
2–4 October 2001	Cancelled due to Foot and Mouth
24–26 September 2002	Ballacolla, Co. Laois
23–25 September 2003	Ballinabrackey, Co. Meath
28–30 September 2004	Tullow, Co. Carlow
27–29 September 2005	Midleton, Co. Cork
27–30 September 2006	Tullow, Co. Carlow

25–27 September 2007	Tullamore, Co. Offaly
23–25 September 2008	Cuffesgrange, Co. Kilkenny
22–24 September 2009	Athy, Co. Kildare
21–23 September 2010	Athy, Co. Kildare
20–22 September 2011	Athy, Co. Kildare
25–27 September 2012	New Ross, Co. Wexford
24–26 September 2013	Ratheniska, Co. Laois
23–25 September 2014	Ratheniska, Co. Laois
22–24 September 2015	Ratheniska, Co. Laois
20–22 September 2016	Screggan, Co. Offaly
19–21 September 2017	Screggan, Co. Offaly

Appendix 2: National Ploughing Champions since 1931

SENIOR HORSE PLOUGH

1931	Ned Jones, Wexford
1932	Michael Redmond, Wexford
1933	Ncd Jones, Wexford
1934	Michael Redmond, Wexford
1935	Hugh Pierce, Wicklow
1936	Michael Redmond, Wexford
1937	Ned Jones, Wexford
1938	David O'Connor, Wexford
1939	Michael Redmond, Wexford
1940	Michael Redmond, Wexford
1941	William Kehoe, Wexford
1942	Michael Redmond, Wexford
1943	Patrick Nolan, Kilkenny
1944	William Kehoe, Wexford
1945	Peter Murphy, Kilkenny
1946	John Halpin, Wicklow
1947	John Halpin, Wicklow
1948	Liam O'Connor, Laois
1949	William Kehoe, Wexford
1950	William Kehoe, Wexford
1951	Michael Redmond, Wexford
1952	Patrick Mahoney, Kerry
1953	Thomas McDonald, Carlow

1954   J. J. Egan, Kerry
1956   Joseph Byrne, Carlow
1957   John Kent, Wexford
1958   J. J. Egan, Kerry
1959   John Halpin, Wicklow (Jan.) and John Halpin,
        Wicklow (Oct.)
1960   Patrick Mahoney, Kerry
1961   J. J. Egan, Kerry
1962   Thomas McDonald, Carlow
1963   Patrick Mahoney, Kerry
1964   Peter Byrne, Wicklow
1965   John Halpin, Wicklow and Peter Byrne, Wicklow (Tie)
1966   Thomas McDonald, Carlow
1967   Thomas McDonald, Carlow
1968   John Halpin, Wicklow
1969   John Kent, Wexford
1970   John Halpin, Wicklow
1971   Jerry Horgan, Cork
1972   Jerry Horgan, Cork
1973   John Kent, Wexford
1974   John Halpin, Wicklow
1975   Jerry Horgan, Cork
1976   John Halpin, Wicklow
1977   J. J. Egan, Kerry
1978   Thomas Reilly, Galway
1979   Michael Kinsella, Wexford
1980   John Halpin, Wicklow
1981   Thomas Reilly, Galway
1982   Thomas Reilly, Galway
1983   Thomas Reilly, Galway
1985   Thomas Reilly, Galway

1986	John Halpin, Wicklow
1987	Thady Kelleher, Cork
1988	Thomas Reilly, Galway
1989	Thomas Reilly, Galway
1990	Joe Fahy, Galway
1991	Thomas Reilly, Galway
1992	Murty Fitzgerald, Cork
1993	Thomas Reilly, Galway
1994	Joe Fahy, Galway
1995	J. J. Delaney, Cork East
1996	Thady Kelleher, Cork East
1997	Thady Kelleher, Cork East
1998	Gerry King, Louth
1999	Thady Kelleher, Cork East
2000	Joe Fahy, Galway
2002	Gerry King, Louth
2003	Gerry King, Louth
2004	Gerry King, Louth
2005	John Sheehan, Cork East
2006	John Sheehan, Cork East
2007	Gerry King, Louth
2008	Gerry King, Louth
2009	Gerry King, Louth
2010	Gerry King, Louth
2011	Gerry King, Louth
2012	Gerry King, Louth
2013	Gerry King, Louth
2014	Gerard Reilly, Galway
2015	Gerry King, Louth
2016	Gerard Reilly, Galway

JUNIOR HORSE PLOUGH

1937	Luke Bracken, Westmeath
1938	Thomas Bambrick, Laois
1939	T. Burke, North Tipperary and J. Hunt, Kildare (Tie)
1940	Thomas Phelan, Waterford
1941	John Joe Egan, Kerry
1942	George O'Brien, Cork
1943	Patrick Maher, Meath
1944	Ed O'Meara, Tipperary and Martin Ward, Galway (Tie)
1945	Thomas O'Dea, Tipperary
1946	John Casey, Offaly
1947	Michael Drennan, Laois
1948	Patrick Mahoney, Kerry
1949	Thomas McDonald, Carlow
1950	Michael Killeen, Tipperary
1951	William R. Murphy, Dublin
1952	William R. Murphy, Dublin
1953	Eugene McGeer, Longford
1954	Thomas McDonald, Carlow
1955	John Murphy, Kerry
1956	Matthew Meleady, Kildare
1957	John Bergin, Tipp. Sth
1958	John Bergin, Tipp. Sth
1959	Henry Campbell, Monaghan (Jan.) and Joseph Scully, Offaly (Oct.)
1960	John Corbett, Clare
1961	Michael Holohan, Kilkenny
1962	Edward O'Sullivan, Waterford
1963	Thomas Reilly, Galway
1964	John Kent, Wexford

1965   M. Behan, Waterford
1966   Ted Keohane, Cork

SENIOR CONVENTIONAL

1949   John Butterley, Louth
1950   John Butterley, Louth
1951   P. J. Shanahan, Kerry
1952   William Woodroofe, Wicklow
1953   Andrew Rodgers, Louth
1954   William R. Murphy, Dublin
1955   William Kehoe, Wexford
1956   Henry Reilly, Louth
1957   Con Slattery, Kerry
1958   William R. Murphy, Dublin
1959   Michael Muldowney, Kilkenny
1960   Richard Mulanny, Kilkenny
1961   William R. Murphy, Dublin
1962   Michael Muldowney, Kilkenny
1963   Michael Muldowney, Kilkenny
1964   Andrew Cullen, Wexford
1965   Michael Muldowney, Kilkenny
1966   Michael Muldowney, Kilkenny
1967   James Murphy, Carlow
1968   William R. Murphy, Dublin
1969   Michael Muldowney, Kilkenny
1970   William R. Murphy, Dublin
1971   Charles Keegan, Wicklow
1972   Willie Ryan, Kilkenny
1973   John Tracey, Carlow
1974   Willie Ryan, Kilkenny

1975	Michael Keating, Wexford
1976	Joseph Shanahan, Wexford
1977	John Tracey, Carlow
1978	Cyril Dineen, Cork
1979	John Somers, Wexford
1980	John Tracey, Carlow
1981	Martin Kehoe, Wexford
1982	William Ryan, Kilkenny
1983	John Tracey, Carlow
1984	John Somers, Wexford
1985	John Tracey, Carlow
1986	Liam Rohan, Laois
1987	Martin Kehoe, Wexford
1988	Martin Kehoe, Wexford
1989	Martin Kehoe, Wexford
1990	Martin Kehoe, Wexford
1991	Martin Kehoe, Wexford
1992	Martin Kehoe, Wexford
1993	Martin Kehoe, Wexford
1994	Martin Kehoe, Wexford
1995	Martin Kehoe, Wexford
1996	Martin Kehoe, Wexford
1997	Martin Kehoe, Wexford
1998	Martin Kehoe, Wexford
1999	Charles Bateman, Cork West
2000	Eamonn Tracey, Carlow
2002	David Wright, Derry
2003	Eamonn Tracey, Carlow
2004	John Tracey, Carlow
2005	Eamonn Tracey, Carlow
2006	Eamonn Tracey, Carlow

2007   William J. Kehoe, Wexford
2008   John Tracey, Carlow
2009   William John Kehoe, Wexford
2010   Eamonn Tracey, Carlow
2011   Eamonn Tracey, Carlow
2012   Eamonn Tracey, Carlow
2013   Eamonn Tracey, Carlow
2014   Eamonn Tracey, Carlow
2015   Eamonn Tracey, Carlow
2016   Eamonn Tracey, Carlow

QUEEN OF THE PLOUGH

1954   Elizabeth Murphy, Louth
1955   Anna Mai Donegan, Kerry
1956   Anna Mai Donegan, Kerry
1957   Muriel Sutton, Wicklow and Eileen Duffy,
       Galway (Tie)
1958   Peggy Mulally, Kilkenny
1959   Mary Shanahan, Kerry (Jan.) and Muriel Sutton,
       Wicklow (Nov.)
1960   Mary Shanahan, Kerry
1961   Eileen Brennan, Laois
1962   Angela Galgey, Waterford
1963   Eileen Brennan, Laois
1964   Eileen Brennan, Laois
1965   Bridget O'Connor, Wexford
1966   Bridget O'Connor, Wexford
1967   Bridget O'Connor, Wexford
1968   Betty Williams, Wicklow
1969   Lillian Keating, Wexford
1970   Mary Ryan, Kilkenny

1971	Maura Murphy, Dublin
1972	Maura Murphy, Dublin
1973	Mary Ryan, Kilkenny
1974	Lillian Stanley, Cork
1975	Majella Ffrench, Wexford
1976	Majella Ffrench, Wexford
1977	Gretta O'Toole, Wicklow
1978	Pauline O'Toole, Wicklow
1979	Teresa McDonald, Carlow
1980	Lillian Stanley, Cork
1981	Alice Murphy, Kilkenny
1982	Teresa McDonald, Carlow
1983	Elizabeth McCaul, Monaghan
1984	Elizabeth McCaul, Monaghan
1985	Marion Stanley, Cork West
1986	Marion Stanley, Cork West
1987	Marion Stanley, Cork West
1988	Elizabeth Lynch, Cork West
1989	Theresa McDonald-Butler, Carlow
1990	Elizabeth Lynch, Cork
1991	Deirdre Barron, Wexford
1992	Marion Stanley, Cork West
1993	Breda Brennan, Carlow
1994	Caroline Dowling, Carlow
1995	Breda Brennan-Murphy, Carlow
1996	Deirdre Barron, Wexford
1997	Fiona Claffey, Offaly
1998	Breda Brennan-Murphy, Carlow
1999	Breda Brennan-Murphy, Carlow
2000	Michelle Kehoe, Wexford
2002	Michelle Kehoe, Wexford

2003	Michelle Kehoe, Wexford
2004	Breda Brennan-Murphy, Carlow
2005	Christine Kehoe, Wexford
2006	Lisa Hartley, Kilkenny
2007	Michelle Kehoe, Wexford
2008	Anna Marie McHugh, Laois
2009	Eleanor Kehoe, Wexford
2010	Bernadette Nolan, Wexford
2011	Joanne Deery, Monaghan
2012	Eleanor Kehoe, Wexford
2013	Joanne Deery, Monaghan
2014	Joanne Deery, Monaghan
2015	Joanne Deery, Monaghan
2016	Laura Grant, Offaly

UNDER-40 HORSE PLOUGH

1990	Gerry Reilly, Galway
1991	Gerry Reilly, Galway
1992	Gerry Reilly, Galway
1993	Tommy Burke, Galway
1994	Pat Fitzgerald, Cork West
1995	Seamus Byrne, Wexford
1996	Zwena McCullough, Cork
1997	Gerry King, Louth
1998	Scamus Byrne, Wexford
1999	James Kilgarriff, Galway
2000	Sean Gilligan, Galway
2002	Gerard King, Louth
2003	Gerard King, Louth
2004	Gerard King, Louth

2005	Gerard King, Louth
2006	Gerard King, Louth
2007	Gerard Collins, Cork West
2008	David King, Louth
2009	David King, Louth
2010	Gerard Collins, Cork West
2011	Gerard Collins, Cork West
2012	David King, Louth
2013	David King, Louth
2014	Gerard Collins, Cork West
2015	David King, Louth
2016	Jeremiah Delaney, Cork East

SPECIAL HORSE PLOUGH

1989	Murty Fitzgerald, Cork
1990	Thomas Reilly, Galway
1991	Joe Fahy, Galway
1992	Thomas Reilly, Galway
1993	Benny Moen, Monaghan
1994	Thady Kelleher, Cork East
1995	Thomas Barrett, Kerry
1996	No competition held
1997	Thady Kelleher, Cork East
1998	Gerry King, Louth
1999	Thady Kelleher, Cork East
2000	Gerry King, Louth
2002	Gerry King, Louth
2003	Gerry King, Louth
2004	Gerry King, Louth
2005	David King, Louth

2007	Gerard King, Louth
2008	Kevin Doran, Wicklow
2009	Gerard King, Louth
2010	James Kilgarrife, Galway
2011	Gerard King, Louth
2012	James Kilgarrife, Galway
2013	Gerard King, Louth
2014	Tim Lawlor, Cork West
2015	Gerard King, Louth
2016	Gerard King, Louth

UNDER-21 CONVENTIONAL PLOUGH

1960	Michael Keating, Wexford
1961	Patrick Brennan, Wicklow
1962	Donal Regan, Kerry
1963	John O'Hea, Cork
1964	Patrick McKeown, Louth
1965	Martin Kehoe, Wexford
1966	John P. O'Donovan, Cork
1967	Eamon Muldowney, Kilkenny
1968	Martin Kehoe, Wexford
1969	Martin Kehoe, Wexford
1970	Seamus Cullen, Wexford
1971	Brian Kieran, Louth
1972	Sean Ffrench, Wexford
1973	Michael Halloran, Kerry
1974	Thomas McDonald, Carlow
1975	Thomas Cuddihy, Kilkenny
1976	Robert O'Connor, Wexford
1977	Frank Gowing, Offaly

1978	Robert O'Connor, Wexford
1979	Robert O'Connor, Wexford
1980	Frank Curtis, Wexford
1981	Jonathan Kehoe, Wexford
1982	Robert O'Connor, Wexford
1983	Frank Curtis, Wexford
1984	Frank Curtis, Wexford
1985	Frank Curtis, Wexford
1986	Mervyn Buttimer, Cork
1987	Eamon Meade, Laois
1988	Richard Stanley, Cork West
1989	Eamonn Tracey, Carlow (Snr) and Harry Mallan, Donegal (Jnr)
1990	See Under-23 Class
1991	See Under-23 Class
1992	D. J. Keohane, Cork (Snr) and G. Prendergast, Tipp. South (Jnr)
1993	Willie J. Kehoe, Wexford (Snr) and Patrick Quinn, Clare (Jnr)
1994	Kevin O'Driscoll, Cork West (Snr) and Anthony Reynolds, Longford (Jnr)
1995	Kevin O'Driscoll, Cork West (Snr) and Justin Mahon, Offaly (Jnr)
1996	Seamus Curtis, Wexford (Snr) and James Coulter, Down (Jnr)
1997	Seamus Curtis, Wexford (Snr) and Garry Simms, Donegal (Jnr)
1998	Kenneth Doyle, Carlow (Snr) and Fergus Courtney, Cork West (Jnr)
1999	Kenneth Doyle, Carlow (Snr) and Alan Davis, Laois (Jnr)

2000 Paul Keating, Wexford (Snr) and Ian Simms,
Antrim (Jnr)

2002 Keith Peoples, Donegal (Snr) and Paul Kirwan, Waterford (Jnr)

2003 Padraig Brandon, Laois (Snr), Thomas Hartley, Kilkenny (Jnr) and Rory Wright, Westmeath (Jnr) (Tie)

2004 Padraig Brandon, Laois

2005 Thomas Hartley, Kilkenny (Snr) and Robert John
Barry, Cork East (Jnr)

2006 Gerard Coakley, Cork West (Snr) and Patrick McCabe,
Monaghan (Jnr)

2007 Sean Tracey, Carlow (Snr) and Sinéad Corbett, Clare
(Jnr)

2008 Sean Tracey, Carlow (Snr) and Patrick McCabe, Monaghan (Jnr)

2009 Kieran Coakley, Cork West (Snr) and Seán
Monaghan, Longford (Jnr)

2010 Kieran Coakley, Cork West (Snr) and Kevin Egan,
Offaly (Jnr)

2011 Sean Tracey, Carlow (Snr) and Lorcan Bergin, Tipp.
South (Jnr)

2012 Sean Tracey, Carlow (Snr) and Mickey O'Halloran,
Kerry (Jnr)

2013 Gerard Kirby, Cork West (Snr) and Lee Simms,
Donegal (Jnr)

2014 Gerard Kirby, Cork West (Snr) and Jamie Hayes, Cork
East (Jnr)

2015 Lee Simms, Donegal (Snr) and P. J. Hartley,
Kilkenny (Jnr)

2016 Lorcan Bergin, Tipp. South (Snr) and Aidan
O'Donovan, Cork West (Jnr)

UNDER-23

1990　Mervyn Buttimer, Cork
1991　Eamon Meade, Laois

UNDER-28 CONVENTIONAL PLOUGH

1964　John Drumm, Kerry
1965　John O'Hea, Cork
1966　John Tracey, Carlow
1967　Anthony Mulligan, Wicklow
1968　Martin Hanion, Wexford
1969　John Paul Donovan, Cork
1970　John Sally, Wicklow
1971　Gerry Coakley, Cork
1972　John O'Driscoll, Cork
1973　Robert Moore, Carlow
1974　Sean Ffrench, Wexford
1975　Robert Moore, Carlow
1976　Martin Kehoe, Wexford
1977　Michael Halloran, Kerry
1978　Nicholas Kelly, Kilkenny
1979　Sean Ffrench, Wexford
1980　Jim Clancy, Cork
1981　Charles Bateman, Cork
1982　Robert O'Connor, Wexford
1983　Jonathan Kehoe, Wexford
1984　Jonathan Kehoe, Wexford
1985　Charles Bateman, Cork
1986　Frank Curtis, Wexford
1987　Charles Bateman, Cork

1988   Joseph Slattery, Tipp. North
1989   Joseph Slattery, Tipp. North
1990   Joseph Slattery, Tipp. North
1991   Colm Dineen, Cork West
1992   Eamonn Tracey, Carlow (Snr) and Harry Mallan, Donegal (Jnr)
1993   Philip Healy, Kerry (Jnr)
1994   Eamonn Tracey, Carlow (Snr) and Enda McKeown, Louth (Jnr)
1995   Eamon Meade, Laois (Snr) and Alan Cox, Offaly (Jnr)
1996   Eamonn Tracey, Carlow (Snr) and Sean Gilligan, Galway (Jnr)
1997   Eamonn Tracey, Carlow (Snr) and Andrew Carnegie, Sligo (Jnr)
1998   Sean Keating, Wexford (Snr) and Wm. Gray, Kildare (Jnr)
1999   Sean Keating, Wexford (Snr) and Patrick Doherty, Donegal (Jnr)
2000   Sean Keating, Wexford (Snr) and Philip Murphy, Waterford (Jnr)
2002   Paul Keating, Wexford (Snr) and Dave Mulcahy, Cork East (Jnr)
2003   Paul Keating, Wexford (Snr) and Alan Davies, Laois (Jnr)
2004   Paul Keating, Wexford
2005   Barry O'Driscoll, Cork West
2006   Martin Kehoe, Wexford
2007   Matthew Simms, Donegal (Snr) and Paul Kirwan, Waterford (Jnr)
2008   Matthew Simms, Donegal (Snr) and Gerard Frost, Clare (Jnr)

2009   Thomas Hartley, Kilkenny (Snr) and Derek
        O'Driscoll, Kerry (Jnr)

2010   Padraig Brandon, Laois (Snr) and Thomas Hartley,
        Kilkenny (Snr) (Tie), Michael John Dillane, Limerick (Jnr)

2011   Padraig Brandon, Laois (Snr) and John O'Brien, Tipp.
        South (Jnr)

2012   Kieran Coakley, Cork West (Snr) and Sean
        Monaghan, Longford (Jnr)

2013   Kieran Coakley, Cork West (Snr) and Cian Keohane,
        Cork East (Jnr)

2014   Brendan Greene, Laois (Snr) and David Owens,
        Kildare (Jnr)

2015   Kieran Coakley, Cork West (Snr) and Frank Cullen,
        Wexford (Jnr)

2016   Kieran Coakley, Cork West (Snr) and Aidan Higgins,
        Galway (Jnr)

THREE-FURROW CONVENTIONAL MATCH PLOUGH

1960   Richard Mullally, Kilkenny
1961   William R. Murphy, Dublin
1962   Joseph T. Hoyne, Offaly
1963   John Walsh, Cork
1964   John O'Hea, Cork
1965   William R. Murphy, Dublin
1966   Patrick Tallon, Meath
1967   Liam Rohan, Laois
1968   Michael Murphy, Carlow
1969   Michael Keating, Wexford
1970   Michael Keating, Wexford

1971 Liam Rohan, Laois

1972 Thomas Phelan, Kilkenny

1973 Dawson Bagnell, Laois

1974 Michael Murphy, Carlow

1975 Michael Tracey, Carlow

1976 Ben Buttimer, Cork

1977 Edward Dowse, Wicklow

1978 Thomas Phelan, Kilkenny

1979 Hugh Reilly, Louth

1980 Hugh Reilly, Louth

1981 Hugh Reilly, Louth

1982 Jim Byrne, Wexford

1983 Michael McEvoy, Laois

1984 Edward Dowse, Wicklow

1985 Hugh Reilly, Louth

1986 Peadar Shortt, Wicklow

1987 D. J O'Driscoll, Cork

1988 Michael Murphy, Carlow

1989 James Byrne, Wexford

1990 Pat Brandon, Laois

1991 Peadar Shortt, Wicklow

1992 Richard Keegan, Carlow (Snr)

1993 Richard Keegan, Carlow (Snr) and Rodney Cox, Offaly (Jnr)

1994 Jerry O'Donovan, Cork West (Snr) and John Dunne, Kildare (Jnr)

1995 Patrick O'Mahony, Cork West (Snr) and Cecil McCullagh, Cavan (Jnr)

1996 Stanley Deane, Cork West (Snr) and Sean Daly, Galway (Jnr)

1997 Rodney Cox, Offaly (Snr) and Kieran McGee, Monaghan (Jnr)

1998   M. L. McEvoy, Laois (Snr) and Peadar Shortt,
        Wicklow (Jnr)
1999   Barry O'Sullivan, Cork West (Snr) and Patrick Beirne,
        Roscommon (Jnr)
2000   Jim Grace, Cork West (Snr) and Gerard Hannon,
        Clare (Jnr)
2002   Richard Keegan, Carlow
2003   John A. O'Donovan, Cork West
2004   Billy Donnelly, Wexford
2005   Richard Keegan, Carlow
2006   James Prendergast, Waterford
2007   Richard Keegan, Carlow
2008   James Prendergast, Waterford
2009   Richard Keegan, Carlow
2010   Pat Furlong, Wexford
2011   Richard Keegan, Carlow
2012   Pat Furlong, Wexford
2013   James Prendergast, Waterford
2014   Pat Furlong, Wexford
2015   Padraig Brandon, Laois

JUNIOR CONVENTIONAL PLOUGH

1960   James Murphy, Carlow
1961   Matthew Dunne, Laois
1962   James Savage, Louth
1963   Christy Grainger, Kildare
1964   Donal Keohane, Cork
1965   Patrick Tallon, Meath
1966   Patrick McKeown, Louth
1967   Patrick McKeown, Louth

1968	Joseph T. Hoyne, Offaly
1969	John Bergin, Tipp. South
1970	Patrick Tallon, Meath
1971	Christy Grainger, Kildare
1972	Gerard Frost, Clare
1973	Gerard Frost, Clare
1974	Roger Callinan, Tipp. North
1975	Thomas Kirwan, Waterford
1976	Thomas Healy, Kerry
1977	Joe McConnon, Monaghan
1978	Larry Bergin, Tipp. South
1979	Matt Duffy, Meath
1980	Philip Murphy, Waterford
1981	Larry Bergin, Tipp. South
1982	Bobby Connolly, Louth
1983	Gerard Frost, Clare
1984	Larry Bergin, Tipp. South
1985	Frank Gowing, Offaly
1986	Matt Duffy, Meath
1987	Christy Grainger, Kildare
1988	Christy Grainger, Kildare
1989	W. P. Buchanan, Donegal
1990	Paul Daly, Westmeath
1991	Bobby Connolly, Longford
1992	Tom Kirwan, Waterford
1993	Larry Bergin, Tipp. South
1994	Kevin Smith, Cavan
1995	Matt Duffy, Meath
1996	Harry Mallon, Donegal
1997	Frank Corbett, Clare
1998	William O'Donnell, Tipp. South

1999	Owen O'Sullivan, Limerick
2000	Tommy Monaghan, Longford
2002	John Donnelly, Wicklow
2003	Frank Corbett, Clare
2004	Alan Cox, Dublin
2005	Gerard Frost, Clare
2006	Dominic Dunne, Kildare
2007	Philip Murphy, Waterford
2008	Michael McEvoy, Laois
2009	Noel Howley, Tipp. South
2010	Christopher Carton, Westmeath
2011	John Molyneaux, Limerick
2012	Owen O'Sullivan, Limerick
2013	Trevor Cobbe, Offaly
2014	Derek O'Driscoll, Kerry
2015	Michael John Dillane, Limerick
2016	Liam Hamilton, Wicklow

STANDARD THREE-FURROW
CONVENTIONAL PLOUGH

1990	Barry Cunningham, Louth
1991	John Dunne, Kildare
1992	John Kirby, Cork West
1993	John Kinsella, Wexford
1994	Noel Jennings, Cork West
1995	Willie O'Donovan, Cork East
1996	Patrick J. Nugent, Waterford
1997	Thomas Grennan, Kilkenny
1998	Noel Walsh, Kilkenny
1999	Cornelius Buckley, Cork West

2000	John Whelan, Wexford
2002	Jonathan Murray, Offaly
2003	Michael Connolly, Offaly
2004	Murt O'Sullivan, Wexford
2005	Pat Furlong, Wexford
2006	Paddy Harrington, Cork East
2007	Jimmy Nyhan, Cork West
2008	John Cottrell, Kilkenny
2009	Jamie O'Sullivan, Cork East
2010	Jim Barrett, Cork East
2011	Niall Twomey, Cork East
2012	Michael Foran, Kilkenny
2013	Patrick Cooley, Wexford
2014	Kieran Dunne, Kildare
2015	Edward Forristal, Kilkenny

THREE FURROW CONVENTIONAL PLOUGH

2016	James Prendergast, Waterford (Snr) and Matthew Coakley, Cork West (Jnr)

REVERSIBLE PLOUGH

1992	Pat Barron, Waterford
1993	James Walsh, Wexford
1994	James Walsh, Wexford
1995	George Murphy, Down
1996	James Walsh, Wexford
1997	Adrian Jameson, Antrim
1998	James Walsh, Wexford (Snr)
1998	Donald Curtis, Wexford (U28)

1999   James Walsh, Wexford (Snr)

1999   Seamus Curtis, Wexford (U28)

2000   Brian Ireland, Kilkenny (Snr)

2000   Con Twomey, Cork East (U28)

2002   Dan Donnelly, Wexford (Snr)

2002   Seamus Curtis, Wexford (U28)

2003   Dan Donnelly, Wexford (Snr)

2003   Declan Buttle, Wexford (U28)

2004   Thomas Cochrane, Dublin (Snr)

2004   Declan Buttle, Wexford (U28)

2005   Brian Ireland, Kilkenny (Snr) and Liam O'Driscoll, Cork West (U28)

2006   John Whelan, Wexford (Snr) and Liam O'Driscoll, Cork West (U28)

2007   John Whelan, Wexford (Snr) and Paul Keating, Wexford (U28)

2008   David Wright, Derry (Snr) and Ger Coakley, Cork West (U28)

2009   Liam O'Driscoll, Cork West (Snr) and Jeremiah Coakley, Cork West (U28)

2010   John Whelan, Wexford (Snr) and Jeremiah Coakley, Cork West (U28)

2011   Jer Coakley, Cork West (Snr) and Paddy Harrington, Cork East (Inter) and Jer Coakley, Cork West (U28)

2012   John Whelan, Wexford (Snr) and Brian Ireland, Kilkenny (Inter) and Martin Kehoe, Wexford (U28)

2013   John Whelan, Wexford (Snr) and Patsy Condron, Laois (Inter) and Jer Coakley, Cork West (U28)

2014   John Whelan, Wexford (Snr) and Enda Kelly, Offaly (Inter) and Jer Coakley, Cork West (U28)

2015   John Whelan, Wexford (Snr) and Tommy McCarthy,
        Kerry (Inter) and Jer Coakley, Cork West (U28)
2016   John Whelan, Wexford (Snr) and Jer Coakley, Cork
        West (Inter) and Dermot Ryan, Offaly (U28)

INTERMEDIATE CONVENTIONAL

1988   John O'Connell, Limerick
1989   Trevor Cobbe, Offaly
1990   Frank Gowing, Offaly
1991   Joe McConnon, Monaghan
1992   Michael Lenihan, Cork East
1993   Michael Houlihan, Tipp. North
1994   Larry Bergin, Tipp. South
1995   Ollie Furey, Galway
1996   Pat Brandon, Laois
1997   David Borland, Meath
1998   Eamonn Traccy, Carlow
1999   Pat McKeown, Louth
2000   Michael Cunningham, Monaghan
2002   P. J. McCaul, Monaghan
2003   Charles McHugh, Donegal
2004   Michael O'Halloran, Kerry
2005   Alan Cox, Dublin
2006   Trevor Cobbe, Offaly
2007   Michael Cunningham, Monaghan
2008   John Slattery, Tipp. North
2009   Pat Brandon, Laois
2010   Trevor Cobbe, Offaly
2011   William John Kehoe, Wexford

2012   Anthony Reynolds, Longford
2013   Padraig Brandon, Laois
2014   Thomas Hartley, Kilkenny
2015   Kevin O'Driscoll, Cork West
2016   John Murphy, Cork West

VINTAGE TWO-FURROW TRAILER PLOUGH

1984   Cecil Wray, Armagh
1985   Andy O'Connell, Cork
1986   Michael Hudson, Wicklow
1987   Gerard King, Louth
1988   Michael Hudson, Wicklow
1989   Bertie Murray, Carlow
1990   Finbar Dulea, Cork
1991   Bertie Murray, Carlow
1992   William Murphy, Wexford
1993   Bertie Murray, Carlow and W. M. Murphy, Wexford (Tie)
1994   Bertie Murray, Carlow
1995   Finbar Dulea, Cork West
1996   Bertie Murray, Carlow
1997   Jimmy Evans, Westmeath
1998   Tom McCracken, Down
1999   John Sullivan, Cork West
2000   Bertie Murray, Carlow
2002   William Murphy, Wexford
2003   Ralph Foster, Fermanagh
2004   William Murphy, Wexford
2005   John O'Sullivan, Cork West
2006   James Evans, Westmeath
2007   Bobby Owen, Kildare

2008	Aeneas Horan, Kerry
2009	William Murphy, Wexford
2010	William Murphy, Wexford
2011	William Murphy, Wexford
2012	Aeneas Horan, Kerry
2013	Liam Prendergast, Waterford
2014	Liam Prendergast, Waterford
2015	Aeneas Horan, Kerry
2016	Tommy Cullen, Wexford

VINTAGE SINGLE-FURROW MOUNTED PLOUGH

1988	P. J. Lynam, Westmeath
1989	Terry Gibson, Longford
1990	Roy Porter, Donegal
1991	P. J. Lynam, Westmeath
1992	P. J. Lynam, Westmeath
1993	P. J. Lynam, Westmeath
1994	Terence Gibson, Longford
1995	P. J. Lynam, Westmeath
1996	Danny Glynn, Laois
1997	Terence Gibson, Longford
1998	Danny Glynn, Galway
1999	Noel Cummins, Kildare
2000	Terence Gibson, Longford
2002	Noel Cummins, Kildare
2003	Bruno McCormack, Meath
2004	P. J. Lynam, Westmeath
2005	P. J. Lynam, Westmeath
2006	William Hayden, Wexford
2007	Gordan Jennings, Cork West

2008	Gordan Jennings, Cork West
2009	Bruno McCormack, Meath
2010	Bruno McCormack, Meath
2011	Bruno McCormack, Meath
2012	Bruno McCormack, Meath
2013	Gordan Jennings, Cork West
2014	Bruno McCormack, Meath
2015	Eoin Buttle, Dublin
2016	Gordan Jennings, Cork West

VINTAGE TWO-FURROW MOUNTED PLOUGH

1991	Terry Gibson, Longford
1992	Michael Tobin, Carlow
1993	Paul Lynch, Cork West
1994	Eric Walsh, Wexford
1995	John F. Keohane, Cork West
1996	John F. Keohane, Cork West
1997	Pat Robinson, Donegal
1998	Timmy Horan, Kerry
1999	George Walsh, Wexford
2000	Maurice Fleming, Cork East
2002	Paul Lynch, Cork West
2003	Paul Lynch, Cork West
2004	Noel Shanahan, Wexford
2005	Maurice Fleming, Cork East
2006	Michael Shanahan, Wexford
2007	Michael Shanahan, Wexford
2008	Michael Shanahan, Wexford
2009	Liam Hamilton, Wicklow

2010	Moss Fleming, Cork East
2011	Joe Kelly, Galway
2012	Aidan Hogan, Tipp. North
2013	John Keohane, Cork West
2014	Ethan Harding, Tipp. North
2015	John McBryde, Offaly
2016	Paddy Doyle, Wicklow

STANDARD THREE-FURROW
REVERSIBLE PLOUGH

2007	Fred Bognall, Laois
2008	Tommy Madigan, Kilkenny
2009	Francis Harney, Wexford
2010	Declan Brennan, Laois
2011	Geoffrey Wycherley, Cork West
2012	Jamie O'Sullivan, Cork East
2013	Ger Lawlor, Cork West
2014	Niall Twomey, Cork West
2015	Niall Twomey, Cork West
2016	Frankie Gowing, Offaly

Appendix 3: NPA Chairmen

Since I started with the Ploughing in 1951, I have worked with the following Chairmen:

John Dowling	Dublin
William Gyves	Wicklow
Batt O'Connor	Kerry
Edward Colton	Offaly
Michael T. Connolly	Wexford
John Savage	Louth
Paul Dunican	Westmeath
Lawrence Sexton	Cork
Gerry Baker	Meath
Liam Murphy	Carlow
James O'Doherty	Kildare
David O'Connor	Wexford
William Dowling	Dublin
Henry McDonald	Wicklow
Aidan Murphy	Dublin
Frank Gowing	Offaly
Michael Hayes	Waterford
James Grainger	Kildare
Thomas Fahey	Galway
Michael Mahon	Offaly
Edmond Hally	Tipperary
James Sutton	Dublin
P. J. Lynam	Westmeath
Denis Keohane	Cork

I want to acknowledge the great help they were to me, always available to advise and assist me in carrying out my duties. I will for evermore appreciate the kindness and respect they showed me.

Appendix 4: National Ploughing Association Directors, 2017/2018

Managing Director   Anna May McHugh, Laois

Asst. Managing Directors   Michael Mahon, Offaly
Eddie Hally, South Tipperary
Anna Marie McHugh, Laois

President   James Sutton, Dublin

Vice-Presidents   Thomas Healy, Kerry
Pat McKeown, Louth

Chairman   Denis Keohane, West Cork

Vice-Chairman   Padraig Nolan, Roscommon

Trustees   Thomas Fahey, Galway
Bernard Kenny, Longford

Directors   PJ Lynam, Westmeath
Patsy Condron, Laois
John Deery, Monaghan
John J Donnelan, Clare
John Dunne, Kildare
Gerard Gilligan, Sligo
Thomas Kirwan, Waterford
John Joe Lawless, Mayo

John McKinney, Donegal
John Molyneaux, Limerick
Dave Mulcahy, Cork East
Daniel O'Dwyer, Kilkenny
John O'Malley, Leitrim
Robert Roe, Wicklow
John Tracey, Carlow
Jim Vogan, Cavan
John Whelan, Wexford
Thomas White, Meath
Joe Slattery, North Tipperary

Life Members:

James Grainger, Kildare
Maurice McEnery, Limerick
Eugene McGerr, Longford
David Rice, Kilkenny

Appendix 5: World Ploughing Contests

1953   Cobourg, Ontario, Canada
1954   Killarney, Co. Kerry, Ireland
1955   Uppsala, Sweden
1956   Shillingford, Oxford, Britain
1957   Peebles, Ohio, USA
1958   Honenheim, Stuttgart, Germany
1959   Armoy, Co. Antrim, Northern Ireland
1960   Tor Mancina, Rome, Italy
1961   Grignon, Paris, France
1962   Dronten, East Flevoland, Netherlands
1963   Caledon, Ontario, Canada
1964   Fuchsenbigl, Niederösterreich, Austria
1965   Ringerike, Buskerud, Norway
1966   No contest held in 1966 for scheduling reasons
1967   Christchurch, New Zealand
1968   Salisbury, Rhodesia
1969   Belgrade, Yugoslavia
1970   Horsens, Denmark
1971   Taunton, Somerset, Britain
1972   Mankato, Minnesota, USA
1973   Wellington Bridge, Co. Wexford, Ireland
1974   Helsinki, Finland
1975   Oshawa, Ontario, Canada
1976   Bjertorp, Vara, Sweden
1977   Flevohof, Biddinghuizen, Netherlands
1978   Wickstadt, near Friedberg, Germany

1979	Limavady, Co. Derry, Northern Ireland
1980	Christchurch, New Zealand
1981	Wellington Bridge, Co. Wexford, Ireland
1982	Longord, Tasmania, Australia
1983	Harare, Zimbabwe
1984	Horncastle, Lincolnshire, Britain
1985	Sdr. Naera, Fyn, Denmark
1986	Olds, Alberta, Canada
1987	Marchfield, Austria
1988	Amana, Iowa, USA
1989	Kleppe, Norway
1990	Zeevolde, Flevoland, Netherlands
1991	Limavady, Co. Derry, Northern Ireland
1992	Albacete, Spain
1993	Vastraby Gard, Helsingborg, Sweden
1994	Outram, Dunedin, New Zealand
1995	Egerton University, Njoro, Kenya
1996	Oak Park, Co. Carlow, Ireland
1997	Geelong, Victoria, Australia
1998	Altheim /Landshut, Germany
1999	Pomacle, Marne, France
2000	Lincoln, Britain
2001	Eskjær, Denmark
2002	Bellechasse, Switzerland
2003	Guelph, Ontario, Canada
2004	Limavady, Co. Derry, Northern Ireland
2005	Prague, Czech Republic
2006	Tullow, Co. Carlow, Ireland
2007	Kaunas, Lithuania
2008	Schloss Grafenegg, Austria
2009	Moravske Toplice, Slovenia
2010	Methven, New Zealand

2011	Lindevad, Sweden
2012	Biograd na Moru, Croatia
2013	Olds, Alberta, Canada
2014	St Jean D'Illac, Bordeaux, France
2015	Vestbo, Hjardemålvej, Denmark
2016	Crockey Hill, York, Britain
2017	Egerton University, Njoro, Kenya

Appendix 6. Irish Competitors in World Ploughing Contests, 1953–2016

Tom McDonnell	1953
Ronald Sheane	1953, 1954
Willie Murphy	1954, 1961, 1963
Michael Muldowney	1960, 1961, 1963
Andrew Cullen	1960, 1964, 1965
Pat Tallon	1962
Charles Keegan	1962, 1964, 1965
James Murphy	1967, 1970, 1971, 1988 (Silver Medal 1970)
Richard Byrne	1969
Cyril Dineen	1969, 1979
Pat McKeown	1970
James Shanahan	1971, 1972, 1974
John Tracey	1972, 1973, 1974, 1980, 1997, 2002, 2003, 2005, 2009 (six times runner-up)
Joseph Shanahan	1973, 1976, 1977
Michael Keating	1975, 1976
Willie Ryan	1975, 1978, 1981, 1983
Michael Murphy	1977
Peter Doyle	1978, 1984, 1990, 1992 (Bronze Medal, 1992)
John Somers	1979, 1980, 1981, 1984 (Bronze Medal, 1980)
Liam Rohan	1982, 1985, 1987, 1988

Martin Kehoe	1982, 1986, 1987, 1989, 1992, 1993, 1994, 1995, 1996, 1998, 1999
Jeremiah Coakley	1983, 1985, 1986
Charles Bateman	1989, 1990, 1995, 2000, 2001
Jackie O'Driscoll	1991, 1994, 1996
Frank Curtis	1993
Mervyn Buttimer	1997
Eamonn Tracey	1998, 1999, 2004, 2006, 2007, 2010, 2011, 2012, 2013, 2014, 2015, 2016 (Gold Medal, 1999; runner-up, 2012, 2016)
John Slattery	2000
Pat Brandon	2001
Sean Keating	2002, 2004
Willie John Kehoe	2003, 2008, 2009, 2010 (Silver Medal, 2008)
John Whelan	2005, 2007, 2008, 2011, 2013, 2014, 2015, 2016
Liam O'Driscoll	2006, 2009, 2010 (Silver & Bronze Medal 2009)
Jeremiah Coakley	2012 (award for Best First Time Competitor)

Irish World Champions

Charles Keegan	1964
Martin Kehoe	1994, 1995, 1999
John Whelan	2013
Eamonn Tracey	2014, 2015

Acknowledgements

Looking back on my life, I have been privileged to meet and work with so many wonderful people. It was impossible to mention everyone in the main text, so here I want to take the opportunity to recognize the many people I encountered along my way.

Of course, you never realize or appreciate school days until you're grown up and look back in time, but they were happy days. It was a great start. From school-going days I understood that it's so important to be involved in what's going on all around you – knowing your neighbours, lending a helping hand and making your surroundings a happier and better place to live. That's how I grew up.

I have very fond memories of being actively involved with our camogie club in Ballylinan, where all the girls would train two evenings a week. I vividly recall the joy after winning and the tears shed when defeated. The comradeship amongst the girls is everlasting – indeed, recently I felt a hand coming into mine and a lady saying, 'Anna May, it's almost fifty-five years since I was on the team playing opposite Ballylinan!' I would like to mention our trainer, the late Michael Dempsey – he was a real gentleman and very patient.

Ballylinan GAA club always ensured that we had the use of the GAA pitch and I have worked closely down through the years with clubs right across the country. I admire the great work the GAA does for the youth of our country.

Having cycled to the Irish Countrywomen's Association meetings in Athy, a few ladies asked: 'Why not have our own guild?' So we formed one in 1964. The two surviving members of that era are Bridie Kaye and myself. Another great community activity in Ballylinan was the drama group, with which I was deeply involved as secretary for many years. And of course our Passion Play came out of the drama group.

Words are inadequate to acknowledge the opportunity NPA founder JJ Bergin gave me when he took me on. I was far from capable – while I had completed a secretarial course, my office experience was practically nil – but he was patient and understanding. Sean O'Farrell, too, gave me a blank canvas to develop my knowledge and understanding of the Ploughing – he was a fantastic mentor.

The Championships grew with the guidance and assistance of so many dedicated supporters. For instance, men who travelled long journeys to judge at parish and county competitions, often in difficult weather. There are so many and I would love to list everyone, but a few come to mind who are no longer with us: the late Jimmy McCarthy, Jack Ryall and Dan Connolly (Cork), Michael Hayes (Waterford), Gerry Baker (Meath), Aidan Murphy and Willie Dowling (Dublin), Pat Nolan (Kilkenny), John Savage (Louth) and Frank Gowing (Offaly).

In some cases, I have worked with four generations in the Ploughing – can you imagine that? – and the ploughing folk of Ireland have never let me down.

Today the NPA Executive is my back-up team and I am deeply appreciative of the members' support – they are Denis Keohane, PJ Lynam, Padraig Nolan, Michael Mahon, Eddie Hally, John Whelan, Tom Fahey, James Sutton.

When I'm at a Ploughing site and see a lorry arriving with tractors and ploughs, I really appreciate the efforts of competitors – some coming from as far as 300 miles away from the venue. They're not doing it for the prize money, but for love of the art of ploughing and the honour of representing their county.

Hundreds will take part year after year in the hope that one day they will be written into the history books as National Champions and join the elite who have competed for their country in the World Ploughing Contest.

Horse-boxes arrive – with teams of horses and a second box for the horse plough. This was how it all started and we still have good competition at national level. I commend the horse ploughmen and women and hope they will continue to showcase the art of horse ploughing.

Thanks also to the plot supervisors and plot stewards who volunteer year after year to take up their posts at the Ploughing, come sunshine, hail, rain or snow. And to all our voluntary workers who look after marking out the various competitions and the exhibition ground. Can you imagine the daunting task of starting with a blank canvas every year? It's not always sunshine when they have to go to the fields and get started, but they do it without a word of complaint.

To all the contractors – many of whom have worked with me for decades – who set up a town in the middle of the countryside with all the amenities a town would have and more. You always make it work – no matter how tight the deadline, how difficult the task, how awkward the project. And when we hit difficult conditions, you never count the cost of putting extra staff on stand-by for the night shift. You just get the job done. Thank you.

To the NPA Control Team that works tirelessly behind the scenes to make sure that the Ploughing runs smoothly throughout the day and throughout the night. You are based in a Portakabin behind the NPA HQ on-site and you never get to enjoy the event, such is your commitment to getting the job done and getting everyone through 'another Ploughing' safely and happily. I have every confidence in you all. Thank you.

Here are some of the other people who have made, or continue to make, a wonderful contribution to the success of the Ploughing:

The ticket-sellers, road stewards and road captains who are up at 4.00 each morning for duty at 5.30 am. Many of these people have never arrived at a National Ploughing site in daylight.

The ladies who give the welcome cup of tea and sandwich at the NPA HQ to our road and field stewards – led by Maureen Healion and Collette Garry for the past twenty-five years, and joined in more recent years by Theresa Mahon, Helen Hally, Marion Fahey and Lil Tracey.

The hospitality team that looks after VIPs and constantly checks when I last had a cup of tea – thank you Breda Hovenden, Ann Lacey and Carmel and Eileen Brennan.

The late Paul Dunican (Westmeath), Carrie Acheson (Tipperary) and Gordon Bradley (Laois) for making all the announcements, and Rosemary Scally who has joined the team.

Our results team, which has the onerous task of compiling results each day under the glare of the media, who want to get them broadcast or printed to a deadline.

Our sponsors, whose generous support is crucial to supporting this great outdoor agricultural event.

A huge thanks to the landowners who welcome the Ploughing on to their lands for a crazy few days and discover a whole new use for their farmland! You are the people who allow us to do what we do and you give us the forum to showcase the best of Ireland year after year.

Thanks also to their neighbours, the local residents, for embracing the excitement of the event despite the inconvenience it can cause occasionally. This is minimized with the assistance of the gardaí, who talk to local residents, businesses and schools.

And that leads me to thanking the chief superintendents and gardaí who do tremendous work for the Championships. Traffic is always a worry pre-ploughing, until we see how the first morning shapes up. Gardaí spend months creating traffic plans with us and driving routes over and over (during and outside of work hours) to make sure that the plans will work. I am massively indebted to the excellent and committed teams who ensure smooth traffic flow and thank the many hundreds of gardaí, from right across districts, who make themselves available for Plough-ing duty – we just could not work without you.

Thanks to the county councils, other county and statutory ser-vices that we work closely with from year to year, including the HSE, Fire Services, Safety Officers and the Order of Malta. Your cooperation is the lifeline of the Ploughing and your willingness to meet, facilitate and work with me and my team down through the years has ensured the success of many ploughing events.

I wish to acknowledge the great support the Ploughing has received from machinery exhibitors (some with us up to six dec-ades), the livestock pedigree breeders, dairy sectors, forestry, plant machinery, motor industry, manufacturers and horticultural com-panies – the range of exhibitors that we now have at the Ploughing is tremendous. Also state agencies, government departments, political parties and local enterprise boards all exhibit at the Plough ing now and bring a status to our event that is greatly appreciated.

I would like to recognize the great people who have given their time voluntarily over the years to coordinate the various side shows at the Ploughing and special interest programmes. These include the Vintage Exhibition and Threshing, the Sheep Shear-ing and Sheep Dog Trials, the Fashion Shows, the Entertainment, the Hunt Chase, the Pony Games, the Meggars, the Livestock Exhibition, the Stabling, and ICA demonstrations. If I started to name names, I would have to go right back to the 1950s, but I need to say that I am very humbled by the hours, days and weeks that you have all put into preparing to make that event 'the best yet' – and it always is.

In recent times the association of Enterprise Ireland and the

Ploughing has opened up huge opportunities for the Ploughing internationally and I am really thankful for that and excited about how this will develop and unfold for years to come.

Thanks to principals and teachers in schools throughout the country for your support. So many of you choose the Ploughing as an educational outing for students and this creates massive opportunities for educational stands and exhibits.

Thank you to the many hotels, train stations and tourist offices nationwide that promote our event through brochure displays. And also to landowners right around the country who permit the NPA to display promotional signs.

Most important, thanks to our patrons who never miss this annual outing. It's an occasion for renewing friendships and doing business.

For many years a core group of 'Ploughing Press' has joined the NPA team in our HQ Hotel, and I honestly think some of their Ploughing stories have been shaped by the tales they have heard in hotel foyers and residents' bars down through the years! These people include the late Sean MacConnell – his determination to get *The Irish Times* to cover the Ploughing and his great satisfaction when it became an annual event for the paper is a special memory. Joe O'Brien, the former RTÉ Agricultural Correspondent, always showcased the best of the Ploughing, even on the rainy days. Ray Ryan, formerly of the *Irish Examiner*, still covers the Ploughing in his retirement and is now the father figure in the press office – always giving very sound advice and better than any library on the history of ploughing. Kieran O'Connor, of WLR FM, who started the craze of sporting heroes coming to the Ploughing. Matt Dempsey, former editor of the *Irish Farmers' Journal* and one of the best public speakers I have ever had the joy of listening to. Damien O'Reilly, RTÉ, the young fella of that gang, who is a very loyal Ploughing supporter.

Nowadays, both the print and broadcast media, local and national, attend the Ploughing for the three days. The *Irish Farmers' Journal* crew is ably led by Justin McCarthy. And we're also delighted to welcome: Darragh McCullough, Margaret Donnelly and all at the *Farming Independent* and *Irish Independent*; Noel Dunne, David Markey and the *Irish Farmers' Monthly* team; Agriland.ie; Mary Kennedy, Mary Fanning and RTÉ's *Nationwide* team; Newstalk – Ivan Yates and George Hook are regulars at the Ploughing; RTÉ's Seán O'Rourke, Ella McSweeney and Frances Shanahan (now retired). And special mention to George Lee and the RTÉ news team, who are first on-site and last to leave.

I have only mentioned a selection of the media, but I want to say a huge thank you for the big part they play in the success of the Ploughing. The coverage is fantastic and they capture the essence of the Ploughing every time and never forget to visit the plough fields.

I particularly want to acknowledge Larry Sheedy's contribution and PR work for the Ploughing for many years. He is always at the end of the phone for advice (including during the writing of this book). Larry's father, the late JJ Sheedy, was involved with the NPA since 1933, as was the late Michael O'Connor, Chief Agricultural Officer for Dublin, and both men were in charge of admission on gates at the second World Ploughing Contest in Killarney in October 1954.

A word of thanks to successive governments and political leaders for recognizing the Ploughing and what we are about and taking the time to visit and bring media attention to our efforts. Thanks also to the Presidents of Ireland for gracing us with their presence down through the years, recognizing this agricultural showcase and bringing a huge sense of pride to the people of rural Ireland.

The Ploughing comes about each year due to the voluntary work of the county ploughing associations, in particular the county chairpersons and secretaries, who are in constant touch with headquarters. I am delighted to name here each ploughing club in the NPA and pay tribute to the part they play in making the National Ploughing Association what it is today.

Counties with county ploughing association only are: Carlow, Cavan, Clare, Dublin, Leitrim, Limerick, Longford, Louth, Mayo, Meath, Monaghan, Roscommon, Sligo, Tipperary North, Westmeath.

And counties with a number of clubs are as follows: *Cork East* (Bartlemy, Watergrasshill, Imokilly, Gortroe, Ballyfeard, Twopothouse, Kilbrin, Banteer); *Cork West* (Macroom, Bandon, Kilbritain, Clogagh, Timoleague, Clonakilty, Cahermore, Kilmeen); *Donegal* (Lennon, Lagan Valley, Finn Valley, Donegal); *Galway* (Athenry, Kiltullagh, Kilconnell, Menlough, Moylough, Caherlistrine, Lackagh, Corrandulla); *Kerry* (Ardfert, Ballyheigue, Ballyduff, Abbeydorney, Causeway); *Kildare* (County Ploughing and North West Kildare Club); *Kilkenny* (Danesfort, Tullogher/Glenmore, Mooncoin, Gowran, Johnstown/Urlingford); *Laois* (Ballylinan and district, South Laois, Portlaoise and district); *Offaly* (Killoughey, Tullamore/Durrow, North Offaly, South Offaly); *Tipperary South* (Dualla, Ballylooby, Poulmucka); *Waterford* (Knockanore, Lismore, Clashmore, East Waterford); *Wexford* (Oilgate, Blackwater, Kilmuckridge, Clongeen, Cusinstown, Ballycullane); *Wicklow* (Rathdown, Roundwood, Enniskerry, Carnew, Barndarrig).

Thanks to everyone I have travelled to the World Ploughing with over many years. The excitement when our competitors are announced as winners – you just cannot describe that feeling. We are very highly regarded by other competing countries and I would like to recognize all my World Ploughing associates, we have made wonderful friendships.

Special thanks to the team that has worked with me in the NPA Headquarters down through the years. Thank you for your absolute commitment to the job, your interest and understanding of the

National Ploughing Association and, above all, your loyalty. My appreciation extends to everyone who has worked from a week to a decade by my side. I am only going to name the current team, led by Louise Brennan, Ann Meaney (in her twenty-first year with the NPA), Mary Clare Brennan, Geraldine Hooban, Ann Siney, Maria Mulhall, Mórag Devins, Mairéad Brennan, and not forgetting Anna and Eileen Brennan who always join in for the last few crazy weeks. And I can't let this acknowledgement pass without mentioning Mary Brennan, who worked with me for thirty-three years.

Thanks to the members of the NPA board of directors, who are always at the end of the phone for advice and are so generous with their time despite the fact that what they do is on a voluntary basis.

Thank you to all the organizations that have presented me with a most fantastic array of awards – I am very humbled.

To my many friends, both in and outside of ploughing, who are always very patient and understanding whenever I'm a late arrival or no-show because of Ploughing business – you are true friends.

To Ciarán Medlar from BDO for making me make the decision to write this book! It has been talked about for so long and this year the annual query from Jimmy Rynhart when he and Deirdre Padian from BDO visited the NPA to wrap up the accounts – 'Did you write the book yet, Anna May?' – just spiralled and you made all the connections, Ciarán. So huge thanks to you for that, for always being at the other end of the phone for any clarifications, and for making the whole process so easy.

To Alison Healy for putting my memories on paper. You were always a pleasure to work with, Alison. You kept me focused and made sure I got to the heart of each story. For your help, your focus and your interest in the detail of my story, I am most grateful. Our timeline was crazy from the start, but you kept to the schedule and we made it! A big thanks also to Rachel Pierce for her stalwart work in editing the book and her astute questions.

ACKNOWLEDGEMENTS

A particular thank you to all the photographers and agencies that supplied photos for the picture sections. For pictures in the second picture section, thanks to Teresa Bergin for the image of JJ Bergin on page 1 and to McHales, Mayo for the image on page 13. Thanks also to Frank Taaffe, Athy, for the historical insights you shared with me for the book.

To Michael McLoughlin and Patricia Deevy from Penguin Ireland for teaching me about publishing and being patient with me throughout the process of writing this book. I had no idea there was so much involved. Thank you for your certainty that I would have enough to say to warrant a book, your belief that people would be interested in reading my story and your willingness to put the Ploughing in your front window.

For my own family, I have so many warm memories. We shared everything, down to our bicycles and cars, and often loaned our shillings to each other. I could not have had a happier childhood.

Thanks to my nieces and nephews, who are very precious to me and whom I don't get to catch up with half as much as I would like to: Eileen's children, Donal James and Padraig Brennan, and Elizabeth Byrne; JJ's children, Elma Dunne, Anna Mai Ryan, and JJ, John Edward and Louise Brennan; and Oliver's children, James, Oliver, Pat and David Brennan.

To my son, DJ, for always giving an objective opinion on this book and for his absolute love of ploughing – you sometimes drive us mad in the office with the detail, but we get it eventually! And a very special word of thanks to my daughter, Anna Marie, for making me accept invitations to do things that I would never have dreamt of doing (including going on *The Meaning of Life* and *The Late Late Show*) and to step up and represent the NPA in places I never thought I would. I would not have finished this book without you, Anna Marie, so a huge thanks for all your help.